MOOD DISORDERS
Toward a New Psychobiology

CRITICAL ISSUES IN PSYCHIATRY
A Series for Clinicians

Series Editor: Sherwyn M. Woods, M.D., Ph.D.
University of Southern California School of Medicine
Los Angeles, California

Recent volumes:

THE INTERFACE BETWEEN THE PSYCHODYNAMIC AND
BEHAVIORAL THERAPIES
Edited by Judd Marmor, M.D., and Sherwyn M. Woods, M.D., Ph.D.

LAW IN THE PRACTICE OF PSYCHIATRY
Seymour L. Halleck, M.D.

NEUROPSYCHIATRIC FEATURES OF MEDICAL DISORDERS
James W. Jefferson, M.D., and John R. Marshall, M.D.

ADULT DEVELOPMENT: A New Dimension in Psychodynamic Theory
and Practice
Calvin A. Colarusso, M.D., and Robert A. Nemiroff, M.D.

SCHIZOPHRENIA
John S. Strauss, M.D., and William T. Carpenter, Jr., M.D.

EXTRAORDINARY DISORDERS OF HUMAN BEHAVIOR
Edited by Claude T. H. Friedmann, M.D., and Robert A. Faguet, M.D.

MARITAL THERAPY: A Combined Psychodynamic–Behavioral Approach
R. Taylor Segraves, M.D., Ph.D.

TREATMENT INTERVENTIONS IN HUMAN SEXUALITY
Edited by Carol C. Nadelson, M.D., and David B. Marcotte, M.D.

CLINICAL PERSPECTIVES ON THE SUPERVISION OF
PSYCHOANALYSIS AND PSYCHOTHERAPY
Edited by Leopold Caligor, Ph.D., Philip M. Bromberg, Ph.D.,
and James D. Meltzer, Ph.D.

MOOD DISORDERS: Toward a New Psychobiology
Peter C. Whybrow, M.D., Hagop S. Akiskal, M.D., and
William T. McKinney, Jr., M.D.

EMERGENCY PSYCHIATRY: Concepts, Methods, and Practice
Edited by Ellen Bassuk, M.D., and Ann W. Birk, Ph.D.

DRUG AND ALCOHOL ABUSE: A Clinical Guide to Diagnosis
and Treatment, Second Edition
Marc A. Schuckit, M.D.

A Continuation Order Plan is available for this series. A continuation order will bring
delivery of each new volume immediately upon publication. Volumes are billed only
upon actual shipment. For further information please contact the publisher.

MOOD DISORDERS
Toward a New Psychobiology

Peter C. Whybrow, M.D.
University of Pennsylvania School of Medicine
Philadelphia, Pennsylvania

Hagop S. Akiskal, M.D.
University of Tennessee College of Medicine and
Baptist Memorial Hospital
Memphis, Tennessee

and

William T. McKinney, Jr., M.D.
University of Wisconsin School of Medicine
Madison, Wisconsin

Plenum Press • New York and London

Library of Congress Cataloging in Publication Data

Whybrow, Peter C.
  Mood disorders.

  (Critical issues in psychiatry)
  Includes bibliographical references and indexes.
   1. Affective disorders. 2. Affective disorders—Physiological aspects. 3. Psycho-
biology. I. Akiskal, Hagop S. II. McKinney, William T. III. Title. IV. Series. [DNLM:
1. Affective Disorders. WM 207 W 629m]
RC537.W49 1984   616.85′27   84-3339
ISBN-13: 978-1-4612-9692-8   e-ISBN-13: 978-1-4613-2729-5
DOI: 10.1007/978-1-4613-2729-5

First Printing—April 1984
Second Printing—August 1985

Albrecht Dürer's engraving Melancholia was first printed in 1514 (Chapter 2, ref. 7). It has been considered an allegory of noble melancholy caused by the astrological influence of Saturn—the planet which rises to dominance in the latter part of the calendar year in the Northern Hemisphere. Dürer himself suffered from episodes of melancholy, and the engraving embodies many of the cardinal features of severe depression as we find them manifest today, 470 years later.

Preface

In this book we present a conceptually integrated approach to disorders of mood. These disorders are defined narrowly as the clinical syndromes of mania and melancholia. The latter is our particular focus, for the simple reason that it is more common and thus more is known about it.

Our approach owes much to Adolf Meyer, who first used the term psychobiology. It was he who emphasized in a practical way the importance of the clinician considering the joint contribution of psychosocial and biological factors in the genesis of mental disorders. However, until the 1960s, our relative ignorance of basic mechanisms that link brain and behavior prevented the development of a genuine psychobiological perspective. Thus Meyer's work was concerned largely with teaching the importance of the personal biography and a consideration of social history in the development of mental disorder. We feel that sufficiently rigorous data have now emerged in psychiatry to permit tentative but real psychobiological integration. Affective illness is probably the most promising area for an attempt at such a synthesis. It is our belief that the theory and clinical practice of psychiatry now can be woven into a coherent theme, integrating insights and evidence generated by the psychodynamic, biological, and behavioral methods; hence in part we review the emerging psychobiology of mood disorders with the hope that it can serve as a generic paradigm for other psychiatric syndromes.

As medical texts go, we suspect this book to be a little unusual; sufficiently so, anyway, that we feel compelled to offer a few observations that will help clarify our intent. First, it is designed to be read as a whole rather than sampled, as might be a collection of essays. This is because what we hope to convey evolves as much from the structure of the book itself as from its parts; each chapter leads to the next. Second, while the book is designed principally for residents, clinicians, and medical students knowledgeable in psychiatry, we hope it will also prove of value to a broad range of persons in the mental health field—including those engaged in basic research but who do not have a clinical background, and clinicians who may have little biological or research knowledge. Hence at times we have been intentionally discursive: exploring the history of a particular idea; touching briefly upon the lives of those who made the seminal contributions to such ideas; or reviewing in comparative depth pertinent neurobiology, concepts of systems, or reasons for classification of phenomena. It is our hope that different readers will always find *some* of the side journeys of interest and that in the end all the elements will be recognized as relevant to our central thesis—that the many parts of our current knowledge of affective illness can be related to each other in a meaningful whole. Finally, we should warn the reader that this is not an exhaustive compendium of ongoing research or the particular treatments employed in the care of those who suffer mood disorder. Again, this is by intent. While we believe the research we discuss to be valid and current, the number of studies emerging each month makes it impossible to cite everything and everybody without destroying ourselves and the reader with details. Similarly, there are many good guides to the rote management of depression and mania—especially the pharmacological management. While methods of treatment and some specifics will be found throughout the book, there is no chapter that tells what to do. In fact, it is our hope that for clinician readers the ideas reviewed will help promote new patterns of care which weave in the thread of their own experience rather than merely adopting ours.

We discovered our mutual interest in an integrated approach to the affective disorders in the early 1970s. Subsequently various writings have been produced (although until this none as a team) that were influenced by our many discussions. In addition to our individual teaching at our parent schools, we have also experimented with "integrated" seminars in various parts of the country. It was the encouraging response to these (and the kind persistence of Hilary Evans at Plenum Press) that led finally to this book.

Peter C. Whybrow
Hagop S. Akiskal
William T. McKinney, Jr.

Acknowledgments

Many persons have contributed to the production of this manuscript, in ways both tangible and intangible.

It is clear to each of us that without the ongoing support of our respective universities,* little would have been achieved. While the demands of everyday practice in the clinic and in the academy have prolonged the effort, without such daily stimulation we would have been poorer in ideas and opportunity. During the drafting of the manuscript the Tennessee Department of Mental Health and Mental Retardation and USPHS grants 05931 and 06147 have helped support Dr. Akiskal. The Wisconsin Psychiatric Research Institute and MH2-1892 have contributed financially to Dr. McKinney's support. The authors owe a special debt to the Josiah Macey, Jr., Foundation Faculty Scholars program, which supported Dr. Whybrow during his year at the Clinical Psychobiology Branch of the National Institute of Mental Health when most of this book was written. Subsequently the specific support of this project has been made possible by the benevolence of Mr. and Mrs. Richard Fowler of Hanover, New Hampshire, and the Dartmouth Medical School.

Many colleagues and friends have commented on the manuscript during its various stages. Dr. Akiskal would particularly like to thank William L. Webb, Jr., M.D., who reviewed some of the parts for which he was responsible.

Christy Wright, Judy Beach, Donna Turner, and Kathy Waterman all helped with the typing of early drafts and the gathering of reference material. Our most profound thanks, however, go to Margo Schworm for her extraordinary diligence and dedication in the preparation and editing of the final manuscript. Because of her exceptional skill with the word processor and Professor Dennis Meadow's kind cooperation in making available his IBM system, the latter stages of the project were completed in record time.

P. C. W.
H. S. A.
W. T. M.

*This book was conceived and written while Dr. Whybrow was a member of the faculty of Dartmouth Medical School. Early in 1984 he left Dartmouth to assume his present position at the University of Pennsylvania.

Contents

II. Elements of Present Knowledge

III. Toward a Synthesis

Overview, History, and Classification of Mood Disorders

In this first section of the book we seek to provide a historical, conceptual, and phenomenological framework to our understanding of mood disorders. Dysthymia, literally a disturbance of the mind or will, is ubiquitous as transient human experience. Extended in time and in dimension, however, it has long been recognized as pathological. The nuance of self-perception and the molding of culture may modify the manifest disturbance to some degree, but to the serious student of behavior a phenomenological core is readily apparent. In our first chapter, we outline those behavioral manifestations that constitute clinical disorder, including information from both the objective and subjective dimension, demographic data, and some teleological and literary reflections.

Subsequently in Chapter 2 we begin with an historical review. Such effort we believe to be more than mere academic exercise for at least these two reasons: (1) description of the nature of mood disorder, regardless of theoretical interpretation, has remained remarkably stable over the centuries and thus we can truly learn from past observation and (2) an awareness of the cultural and philosophical underpinning of the explanatory models which have emerged is essential for they covertly influence much of our current research and clinical practice. There remains, for example, a tension between physical and psychological (or spiritual) interpretation which pervades the models that currently seek to explain mood dysfunction. A review of these models, drawing upon the historical survey, completes the second chapter.

In Chapter 3 we take up the issue of classification. Why classify at all? It is a question worthy of discussion especially in medicine, since diagnosis (a form of classification) must carry the burden of predicting outcome. Ideally any classification is based upon an intimate understanding of pathogenesis and prognosis, but in the absence of such knowledge care must be taken not to confuse opinion and observation. This dilemma is discussed at length, and methods of clustering information into meaningful categories are reviewed. Finally the practical usefulness and limitations of such systems are illustrated through reference to the *DSM-III* axial classification of affective illness.

Mood Disorders: An Introduction

There is a pitch of unhappiness so great that the goods of nature may be entirely forgotten, and all sentiment of their existence vanish from the mental field. For this extremity of passion to be reached, something more is needed than the observation of life and reflection upon death. The individual must in his own person become the prey of pathological melancholy. . . . Such sensitiveness and susceptibility of mental pain is a rare occurrence where the nervous constitution is entirely normal: one seldom finds it in a healthy subject even where he is the victim of the most atrocious cruelties of outward fortune . . . it is a positive and active anguish, a sort of psychical neuralgia wholly unknown to healthy life.

William James[1]
The Varieties of Religious Experience (1923)

Two Case Histories

Melancholia

By all objective accounts he was indeed a prominent lawyer. The policeman, who had found him sitting at 4 A.M. in an empty public square chewing aspirin tablets, said he was, and his wife, when called, confirmed it. And yet he described himself as a "shell of a person" fit only to be prosecuted for moral decay. A loud voice—"heh"—was ridiculing him. He was not a man. Had he not neglected his wife, both emotionally and sexually? Was he not cursed and ignored by God as a failure? All his benevolent acts were a cover for self-aggrandizement. He was the model for the wretched lawyer whom Camus had protrayed in his novel *The Fall*. He had indulged himself in a life of pseudoservice. He should never have entered the legal profession. Rather, he should have been a janitor. In fact, he wondered at times whether he *was* a lawyer. Perhaps he had been once, but now he was clearly an imposter, an empty, useless wretch who could no longer concentrate or make the simplest decisions, who paced his office late into the night, afraid to go home and face another restless, sleepless night. He felt caught in something repetitious and uncontrollable, beset by sinful thoughts that threatened to destroy him. He was "wiped out," "fallen apart," so alienated from his creator and his family that he wondered whether he even existed. Who were the authorities to deprive him of his final, perhaps only, courageous act?

He had left his office just before dawn. There had been some planning. He knew that enough aspirin would kill him. It would be a logical solution, a precise reversal of his birth half a century before. He sat down in his favorite public square—on the steps of the Gallery of Art, a place where he had spent so many happy free hours. The colors in the paintings were now pale and drab, drained of their previous ability to provide him with pleasure. So, with fixed stare, preoccupied with thoughts of a wasted life, the events of which now paraded before him in sad contrast to the beauty of the dawn, he proceeded to deliberately and slowly chew the small, bitter white tablets. As the minutes passed, there came a distinct ringing in his ears. It reminded him of his childhood, of mass in the early morning at his father's church.

Approximately 1 in 20 people will experience melancholia during their lifetime.[2] It is an ancient malady, described in the earliest writings of man's experience. It is the subjective state that the Greeks equated with a dark mood, a black bilious humor. It must be distinguished from the "depression" at its fringe, which is accepted as commonplace and ubiquitous. All of us have known sadness and self-doubt, but the destructive power with which the lawyer struggled is an experience peculiar to severe depressive illness.

Mania

In similar mode, a great joy and pleasure in the surrounding world can grow in some to a disruptive euphoria and a devouring excitement. Aretaeus described such a condition in Rome during the second century. The experiences he drew upon probably differed little from the subjective report that follows. The author is a young man, a clerk in a northeastern tax office, frustrated by his work and desperately wishing to be a writer.

"It was in the spring that I first began to be plagued by sleeplessness for which I sought medical advice and was given at various times Elavil and Librium (I think). I was drinking, not heavily but steadily, both socially and in order to help the sleeping problem. It also served to break the boredom of my job. For the first time I felt it was becoming unmanageable. I disliked the sycophancy demanded of the workers—disliked is not strong enough here—I despised any expression of authority, whether it applied to me or not. I was becoming angry over little things, feeling I could run the place better, more efficiently, more humanely. Everything seemed static, futile, even though I was earning a reasonable wage and intended to leave for Europe within weeks.

"Retrospectively I suspect I was a little depressed. I know I was anxious about not sleeping, as I kept telling people about it, wanting to acquaint them with my difficulties. Then during the week or two before departing for Europe, my mood lightened quite suddenly. Things seemed almost humorous at the office and my colleagues like so many robots. I believe I was increasingly anxious to get everything under way, growing tense waiting to be in Europe and relying on more than the prescribed dose of Elavil in order to gain some calm. One evening after having only two glasses of beer (I remember the amount precisely), I sat on the corner of a street downtown and found myself laughing and crying simultaneously. I was struck by the ludicrous nature of the simple actions performed by other people: looking both ways before crossing the road, for instance. People tended to ignore me, and this made the entire situation all the more comical. I began to feel a great sense of energy and a wish to move. I began to defy the traffic, running in and out (I had a sense of fear, but also power which seemed to allow me to take enormous risks). The whole street seemed brilliantly lit, as though from an arc lamp, and I felt wonderful and yelled epithets at the motorists, who stopped and screamed at me.

"Also my increased strength was not imaginary. After I got back to my apartment, I smashed down the kitchen door, although it was locked and although I had the keys with me. I simply kept butting it with my shoulder until I tore it from its hinges. I didn't sleep much, if at all, that night, feeling a great pressure and a sense of elation. I decided to leave for Europe immediately rather than in a week's time as planned and indeed did so only two days later.

"The laughing episode did not return, but I went to Europe with a good deal of energy and buoyancy. I was not there long (a week or two) when I became irascible and difficult to live with. (I had met a girl on the plane and we were touring together.) I felt that she didn't appreciate me, either for my sexual prowess or for my wit. Anyway, I felt I had an idea for a novel and it was imperative for me to return home to line up a publisher.

"I think I remember these episodes in retrospect with more coherence than they had at the time. Some things are just lost. I remember being shocked later, for example, how much money I had spent, even though my stay in Europe had been but a few days. On my return I was trying to write, something I had always wanted to do.

"The work I produced was appalling, however—fragmented, no discipline, sporadic, and uneven. I felt 'other worldly' and under pressure, aware that something must give but convinced I was a genius. There were also short periods of calm, an overwhelming sense of euphoria that was rather nice really, almost mystical. But my emotions weren't all happy, they seemed to fluctuate tremendously: self pity, hatred (I was probably capable of doing physical injury), a diffuse and general love, abject helplessness, hopelessness, and guilt. These last three began to predominate, even though for a time it seemed that whenever I hit a record 'low' some bright and incredible idea came to offer hope temporarily. But decline from each of these plateaus left me deeper in despair, with less capacity to halt the downward trend even temporarily. My inability to confront people reached a pitch. There was a sinking of all my faculties."

The Semantics of Mood and Its Disorder

The uncanny similarity between ancient and modern descriptions of these profound disturbances of mood serves to emphasize the circumscribed nature of their pathology. This is in contrast to marked differences in theories of causation which have emerged over the centuries. Retrospectively, the sufferer usually sees the experience as alien—radically different from the usual self—and the listener has little doubt that what is being described represents an evil or an illness. However, the elements of the description are usually highly personal and the language used is colored by the prevailing mood. That the experiences recounted shade imperceptibly into the suffering of everyday life has long confounded attempts to understand and explain these phenomena. The triggers to subjective awareness of feeling and mood are specific and peculiar to our individual experience. To reduce semantic complexities, research workers and clinicians have increasingly sought operational definitions that have a reasonable degree of agreement between individuals.

Later we shall discuss the classification of mood disorders in some detail, but some immediate clarification of our own terminology will be helpful. Throughout this book, the terms "mood disorder," "affective disorder," and "affective illness" will be used interchangeably. Mania and its milder or nonpsychotic form, known as hypomania, are defined as mood disorders in which elation, excitement, and acceleration predominate. The term "depression," when unqualified, will be used in a generic sense to refer to all varieties of depressive phenomena, from "normal" everyday sadness to depressions of pathological depth or duration such as melancholia. The last term is reserved for the fully developed syndrome of depression in which somatic concomitants dominate the clinical picture. This illness most commonly takes a unipolar form, characterized by one or more episodes of clinical depression. Bipolar or manic–depressive illness, which is less common, is characterized by the occurrence of at least one excited episode.

The terms "melancholia" and "mania" were introduced by the ancient Greeks. Kraepelin, the German psychiatrist who did much to order psychiatric nosology, introduced the term depression in his description of alternating patterns of mania and melancholia under the title of "manic–depressive insanity."[3] Later, largely because of the teachings of Adolph Meyer, the term melancholia was dropped, particularly from the American literature, in favor of the term "depressive reaction." Meyer popularized a "common sense" version of dynamic psychiatry[4] after his emigration to the United States from Switzerland at the turn of the century. He argued that depression, a reaction to a psychologically understandable series of personal reverses, was more appropriately designated as such. In attempting to emphasize the importance of environment, he rejected the "humoral" (biological) implications of the term melancholia.

This element of Meyer's legacy is partially responsible for some of our contemporary clinical and semantic confusion. However, modern practical therapeutics dictate that normal sadness and depressive symptoms that are part of a nonaffective disorder be distinguished from the syndrome of depressive *illness* in which depression of mood is a primary disturbance. Hence recently in *DSM-III,* the third edition of the *Diagnostic and Statistical Manual of the American Psychiatric Association,* we have seen a reintroduction of the term melancholia. We shall return to these important clinical distinctions in Chapter 3.

The Clinical Manifestations of Mood Disorder

On admission to hospital, the prominent lawyer whose experiences were described earlier denied being depressed. Rather, he complained of having "no feelings of any sort. . . . I have no soul, I am dead inside." When pressed, he confided that he believed he suffered from a case of "moral decay of the soul—sin sickness," as he termed it during a flash of his old courtroom eloquence. "The sentence should be electrocution rather than shock treatments." However, after receiving the latter, he no longer believed he deserved to be electrocuted; indeed, in six weeks he was able to return to the practice of law.

It is often difficult to differentiate the illness from the person in the early stages of an episode of affective disorder. Patients themselves, friends and relatives, physicians, and mental health professionals all experience this difficulty. However, at some point, especially upon reflection, it is obvious that a profound change has occurred and the "person" has been temporarily submerged by pathological distortions of behavior and perception. Both the lawyer and the young would-be writer who were described previously were aware in retrospect of the distortion that had taken place in their subjective interpretation of their world. There is a change in the perception of both life events and the self—the thinking self as well as the physical being.

Some of the best descriptions of these disorders are found in Emil Kraepelin's book, *Manic–Depressive Insanity and Paranoia,*[3] first published in English translation in 1921. The interested reader is referred to that text, for here we shall only summarize the phenomena most often associated with mood disorders. For convenience we have chosen to catalogue these in five groups; they are summarized in Table 1.1.

That there is a disturbance of *Mood* is, of course, a unifying concept in the affective disorders. Although the subjective feelings described and expressed by most melancholic and manic individuals do bear some resemblance to the mood changes of everyday life, they clearly go beyond the common experience. Most of the discussion in the literature has focused on the mood typically found in severe clinical depression or melancholia. Happy people do not complain; similarly, the manic rarely speaks spontaneously of mood; rather he or she lives the mood, and it is for others to experience and frequently complain! In contrast, the melancholic individual experiences painful affects and has an inability to respond to or to generate pleasurable stimuli. Approximately 50% of such individuals spontaneously report an overwhelming sadness or depression as their principal complaint. Crying occurs, but as the illness progresses the tears dry up, a phenomenon that usually causes great concern and bewilderment in the individual who experiences it for the first time. It is also important to remember that 10%–15% of melancholic individuals deny sad feelings. Such was the experience of the lawyer. The affective disturbance is so severe and pervasive in such cases that it inhibits the experience of all types of emotion, including sadness. Hence the absence of depressed mood does not exclude melancholia when other signs are present.[5,6]

The manic, by contrast, exudes an infectious gaiety, with much laughter, expansive gestures, and occasional statements which leave no doubt that the individual

Table 1.1 Clinical Manifestations of Mood Disorder

	Melancholia	Mania
Mood	No laughing	Euphoric, much laughter
	Anxious; irritable	Expansive
	Sad; crying (no tears as illness progresses)	Irritable
	Anhedonia	Argumentative
		Hostile
		Momentary tearfulness
Mentation	Poor concentration	Poor concentration
	Indecisive	Distractible, flight of ideas
	Memory impaired	Rhyming and punning
	Helplessness; hopelessness	Grandiose ideas and delusions
	Habit dominates thought; active ruminative thoughts	Memory distortion
	Self-depreciation and guilt	Fleeting hallucinations
	Pessimism, nihilism	
	Delusions of inadequacy, illness, poverty, sin, crime, and punishment	
	Accusatory voices	
	Suicidal thoughts	
Drive activity and behavior	Loss of motivation	Intrusive
	Decreased efficiency	Increased social drive
	Decreased performance, increased error	Sexual drive and interest increased
	Loss of gratification in effort	Uninhibited
	Sexual drive decreased	Inappropriate interpersonal communication with friends and strangers
	Social withdrawal, loss of attachment	Easily triggered to anger and even rage
	Self-destructive acts	Excessive expenditure of money
		Disorganized execution of daily tasks
		Activities disorganized
Somatic manifestation	Fatigue	Increased energy
	Appetite decreased	(Increased strength?)
	Constipation	Appetite increased (although food taken irregularly)
	Aches and pains	May eat bizarre things
	Weight falls	Weight falls
	Sleep decreased (in bipolar illness sleep may increase during depressive phase)	Sleep decreased and irregular; periods of 24–36 hours without sleep
Appearance	Skin pale and dry	Smiling, mobile facies
	Appears preoccupied	Elastic, vigorous movement
	Furrowed face, eyes staring	Flamboyant dress, but often disheveled
	Disheveled, drab dress	
	Poor personal hygiene	
	Stooped posture	

believes life is worth living. However, irritability is often very near the surface, especially when the manic individual is closely questioned; conversations can degenerate rapidly into argument. This irritability is almost as classical in mania as is the much cited euphoria. The irritability is labile—momentary tearfulness and even suicidal ideation not infrequently mar the manic's otherwise joyous disposition. These mixed states[7] provide psychodynamic and biological links between depression and elation and suggest a common psychopathological process.

In both states there is a clear change in thinking or mentation. Both manic and melancholic individuals experience poor concentration, although the former rarely admit this. However, for the depressive it may become a preoccupying and presenting symptom. Depression also steals from the individual the power of memory and decision. Events become jumbled, habit begins to dominate thought, and the individual falls back upon routine, ruminating about the minutiae of life. For some this pattern becomes a signal of impending madness which feeds the nihilism and pessimism that they adopt toward their capabilities and their personal relationships. These cognitive changes[8] can progress to the point where the individual expresses delusional convictions of being ill, poverty-stricken, worthless, and sinful, and may lead, as in the case of the lawyer, to intrusive thoughts of suicide. Others may believe that their failures and moral crimes are of such gravity that they deserve prosecution and punishment; hallucinatory voices accuse and ridicule them. The patients' difficulties are also apparent to friends and relatives. Although these are at first subtle, the irritability and difficulty in collaborating in what were previously mutual tasks produce anger and resentment in family members, thereby compounding the delusional trends in the direction of persecutory thoughts.

Objective measurement shows that mental performance also declines in mania. However, the manic individual may believe that his or her ability to think and reason has never been better. Such was the experience of the Boston tax clerk. Highly distractible, in the early phases of mania the individual is witty, humorous, and delightful to listen to. Ideas run nonstop, frequently laced with a cunning, scheming effort to control large resources or people. As the disturbance progresses, genuine grandiosity appears and the manic's friends and associates begin to become uncomfortable and attempt to avoid his or her company. Memory distortion also occurs, and the manic frequently "reorganizes" his or her recall of the episode with greater skill than that with which the behavior was actually organized at the time.

The *drive* and *activity* of the manic are perhaps the most striking feature of this illness. They are excessive in every way, and as the illness progresses the disorganization, lack of sleep, excessive expenditure of money and other resources, and intrusion into others' lives become so clearly imbedded in the minds of those who live with the manic that they become apprehensive of further attacks. Sexual thoughts and references abound in conversation and rapidly progress to inappropriate and intrusive social behavior. Indeed, the intrusive nature of the manic's activity frequently leads to the breakup of social relationships and is one explanation for the very high divorce rate among individuals who struggle with this group of illnesses.

Living with the depressive is no less trying, especially if there has previously been a close bond. Deficient motivation, diminished initiative, decreased efficiency, and increased error in performance, as well as the loss of self-gratification and the abhorrence of social situations, all combine to foster the melancholic withdrawal from friends and family.

The *vegetative* or *somatic* manifestations may be inferred from what has already been outlined. While the melancholic experiences enormous fatigue, the manic has boundless energy and increased physical strength. Certainly that was the subjective impression and perhaps reality for the young writer who broke down his kitchen door and defied the traffic on a busy street. Appetite also is changed. Whereas food is of little interest to the melancholic, most manics find an increased appetite. However, they eat rather irregularly and frequently eat bizarre things. One patient we recall ate (in small quantities) the poisonous mushroom amanita phalloides to prove—in typical psychotic grandiosity—his immunity to death.

The manic individual often loses weight, but weight may increase during the depressive phase of the bipolar cycle (possibly from inactivity). In the unipolar depressive (where mania does not form part of the illness), however, weight is lost rapidly as the melancholia restricts all behaviors, including eating.

Sleep also differs in bipolar and unipolar melancholic illness, being increased during the depressive phase of manic–depressive illness. In the unipolar melancholic, sleep is disrupted, very poor in quality, and short in length. The manic appears to have a reduced need for sleep, and the illness is characterized by runs of activity, 24–36 hours in length, during which the individual apparently experiences little or no fatigue.

In *appearance* the depressive is pale, disheveled and drab, shrivelled, and preoccupied, with furrowed brow and face, stooped posture, and inadequate personal hygiene. This picture is in marked contrast to the typical manic—a smiling, bubbly, and elastic individual who bounds into a room vigorously inquiring about everybody and everything. The manic's dress is usually flamboyant and may be bizarre, with many personal articles attached to the clothing or carried. In many cases there is a certain dishevelment and poor attention to the personal toilet, however. This extravagant picture is not always present, but on close examination there is frequently an unusual feature that provides a clue. A special example is captured in a photograph in Kraepelin's book *Manic–Depressive Insanity and Paranoia*,[3] in which a rotund smiling individual sits contentedly in a chair smoking both a cigar and a pipe.

Most sufferers do not have all of the clinical features outlined here. Individuals who are prone to recurrent illness often have a pattern of signs which they learn to recognize as heralding another episode of mood disturbance. Similarly, changes such as a return of energy, decreased need for sleep, the realization that a meal has tasted good, and an awareness that one is actually looking forward to the day ahead are recognized through experience as signals of a return to health or the transition to hypomania (see Fig. 1.1).

An important feature of the affective disorders is their episodic nature. When episodes recur, they frequently do so at regular intervals; this is particularly true for bipolar mood swings. Within an illness period, a diurnal pattern is also common, especially in melancholia, in which the symptoms dominate the sufferer's early morning but loosen their grip somewhat as the day advances. Kraepelin[3] noted that in bipolar patients the switch from one phase of illness to another commonly occurs in the early morning. For instance, a patient suffering from a retarded depression may wake up early in the morning after a short night of sleep feeling full of pep and bright spirit.

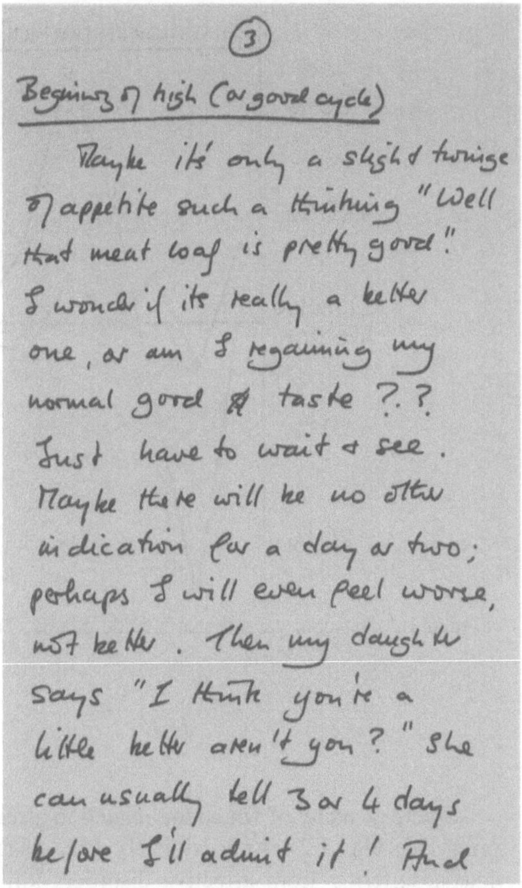

(3)

Beginning of high (or good cycle)

Maybe it's only a slight twinge of appetite such a thinking "Well that meat loaf is pretty good!" I wonder if its really a better one, or am I regaining my normal good taste ?? Just have to wait & see. Maybe there will be no other indication for a day or two; perhaps I will even feel worse, not better. Then my daughter says "I think you're a little better aren't you?" She can usually tell 3 or 4 days before I'll admit it! And

Figure 1.1 Description of subjective experience of switch from depression to normal mood. Taken from original notes made by 67-year-old woman.

Most depressive episodes end within 3–12 months, depending on whether treatment intervenes. Although psychosocial and personality sequelae are not uncommon,[9] *symptomatic* chronicity extending beyond 2 years probably does not occur in more than 10%–15% of all cases.

Manic episodes are even shorter than depressive episodes, and residual chronicity is very rare. The interval between episodes of depression may span many years in unipolar illness, often decades, with somewhat shorter intervals in bipolar illness. Frequently the periods of illness show a circannual pattern. Some individuals, for example, experience a prolonged depressive dip in September and a brief manic episode in April; other patients may experience yearly manic episodes in the fall. The peaks of disturbance do appear to coincide with the spring and autumnal equinox in temperate climates. One female patient recalled by the authors would fail her college courses during the fall quarter but would then more than catch up—indeed excel—during the winter quarter. There are well-documented records of patients who have alternating episodes of depression and mania in cycles as short as 48 hours; pure depressive cycles of short duration have also been reported, but are definitely less common.[10]

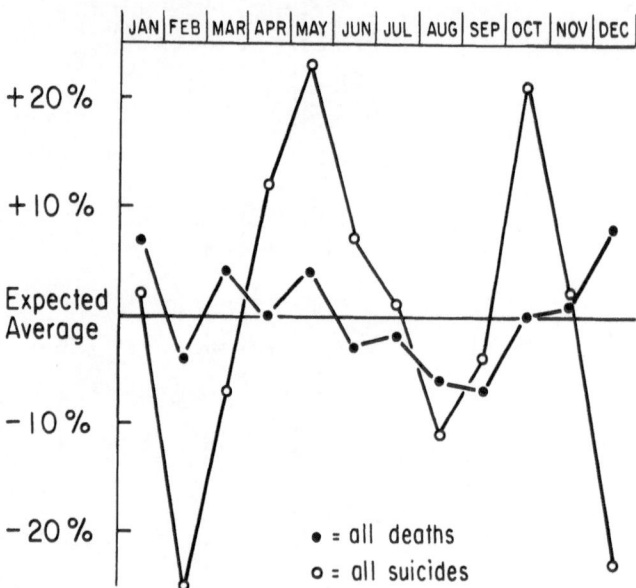

Figure 1.2 Seasonal variation of suicide deaths. Taken from northern New England data.

Suicides, many of which are related to affective illness, also peak during fall and spring (see Fig. 1.2). Approximately 15% of untreated or inadequately treated individuals suffering from affective disorders kill themselves. Studies by Robins and Murphy in the United States and Barraclough et. al. in Great Britain have shown that most completed suicides occur as a result of mental illness, principally depression. [11-13] Affective illness accounts for two-thirds of all suicides; most of the remaining are due to alcoholism. Suicide ranks 10th as a cause of death in the United States for all ages, and is the leading cause of death in young adults.

For approximately one-third of individuals who suffer clinical depression, the episode occurs only once in a lifetime, suggesting that an unusual combination of events has conspired in producing an isolated vulnerability to the disorder in these individuals. Indeed, melancholia is probably best viewed as a final common path of many variables—some inherited, others acquired, but none of which alone is a necessary or sufficient cause. It is these elements clustering in time that "permit" the expression of the melancholic disorder. We may term them permissive factors. By contrast, manic states in recurrent bipolar illness are probably more dependent upon an enduring biological (genetic–biochemical) permission, played upon by environmental precipitants, both psychological and physical. Specific biological factors, such as the use of the steroids or thyroid hormones, however, can lead to isolated episodes of hypomania.

Who Suffers Mood Disorders?

Depression, broadly defined, is not only the most prevalent psychiatric condition in untreated individuals, but also the most common diagnosis in outpatients and hospitalized people—whether in psychiatric or medical settings. Estimates vary from about 20% of diagnoses in state mental hosptials to 40%–50% in outpatient psychiatric facilities, and up to 70% of psychiatric diagnoses made in medical settings.[14]

Current estimates indicate that 15% of all women and 10% of all men will suffer clinical depression during their lifetime. These prevalence figures are relatively conservative and are based on Helgason's study in Iceland,[2] which reported specifically severe depression. Through case-finding in the community—where many untreated and milder cases are identified—recent epidemiological studies conducted by Weissman and Myers[15] in New Haven, Connecticut, suggest a prevalence twice that of the Iceland study.

Otto Fenichel,[16] a psychoanalyst, has suggested that depression represents an appropriate response to certain life circumstances. Similarly, others have commented that depression is such an integral part of the human condition, especially in today's world, that the majority of the population will experience it. Available data, however, do not warrant such pessimism. For instance, Woodruff and associates at Washington University in St. Louis have demonstrated that sustained moods of despondency do not predominate, even in the biological kin of individuals with psychiatric disease—a group at high risk for mood disturbance.[17] Thus, although sadness and transient depressive states are common, perhaps ubiquitous, in human experience, melancholic depression is not. Only when a constellation of special "permissive factors" exists does the major clinical disorder develop; for each individual that is the unique combination in time of environmental, cultural, intrapsychic, and developmental/biological elements.

The popular suggestion that we have entered an "age of depression" appears to have little validity. The lifetime risk of the disabling and more *recurrent* forms of affective illness, as far as can be determined, has probably changed little and remains at no more than 4%–5%; for bipolar illness the figure is lower and stable at about 1%.[15]

The availability of relatively specific forms of therapy may, however, have increased the number of *diagnoses* of affective disorder in recent years. There is evidence that physicians frequently give precedence to a diagnosis for which treatment is available or that holds the promise of a good prognosis. For instance, in the United States, the era of psychotherapy and antipsychotic drug treatments for schizophrenia fostered an increase in that diagnosis relative to affective disorder. Recently, with the introduction of lithium carbonate therapy, the rediscovery that affective illness is self-limiting, and the reduction in stigma that has resulted from improved public education, there has been an increase in the number of affective disorders diagnosed.[18,19] However, we doubt this represents a real increase in the severe forms of the illness.

Whether one considers "mild" depressions or the more profound melancholic syndrome, it does appear that mood disorders are more prevalent in women than in men, at least in our Western culture.[20] Nor does this seem to be merely a bias introduced by "chauvinistic" male psychiatrists. Community surveys keyed to symptomatology have revealed a preponderance of mood disturbance in women, even though many of them have never come to clinical attention. In the absence of evidence that women are more likely than men to report such subjective distress, one must conclude that this is a true epidemiological finding.

In most Western countries, the female-to-male ratio is about 2 : 1 for depressive conditions (and close to even for manic–depressive states). However, this ratio varies in other cultures. Strikingly, it was found in one recent study that women in three Nigerian villages reported a level of depressive symptomatology twice that of housewives in a London suburb.[21] Leighton and co-workers have also shown that depressive symptoms were four times more common in their Nigerian probands than among the population of Stirling County, Canada.[22] Both African samples reported high levels of subjective anergia, self-depreciation, and guilt. Many Western observers had predicted the opposite, operating on the assumption that guilt is a product of Western cultural heritage. Only in India do studies suggest a male preponderance of depressive illness, which, in the absence of community surveys, has been ascribed to preferential utilization of psychiatric services by men. Hence the issue of who is *treated* for depressive illness does, in part, seem to be culturally bound, although the core symptoms of melancholia have been observed in every culture surveyed.

Another consistent demographic characteristic of affective illness is correlation with age. The median age of onset of the more common unipolar disorders is in the late 30s; by contrast, the median age of onset for bipolar disorder is the late 20s, with a large number of cases beginning in adolescence. Recent studies by Hudgens from Washington University and Carlson and Strober from the University of California at Los Angeles have documented that affective disorders are quite common during adolescence and may actually outnumber schizophrenic psychoses.[23,24] At the other end of the life cycle, the affective syndromes are also more common than is generally realized, and are frequently complicated by impaired cognition from pseudodementia.[25]

Evidence suggests that social class and race do not make differential contributions to the overall incidence of depression. Recent studies have not replicated the Hollingshead–Redlich finding which suggested a higher rate of affective illness in the upper social classes.[26] Individuals from a lower socioeconomic group do seem to present with a different symptom cluster, however, with marked symptoms of irritability, anxiety, and somatic discomfort. This may account for earlier reports of a lower rate of affective disorder in these classes. Similarly, claims of lower rates of manic–depressive illness in black patients are not supported by current evidence and may again be explained by overdiagnosis of schizophrenia resulting from the tendency of black patients to present symptoms that the physician perceives as floridly psychotic.[27,28]

The Distinction between Mood Change and Clinical Episode

A state of mood *disorder* implies that one emotion becomes so compelling and pervasive that it has eclipsed rational consideration of other intrapsychic events and of the environment. Mood *change*, however, is for all of us a central component of the kaleidoscope of subjective life. Identifying the precise point of transition from what we choose to define as a dimension of normal mood and that which is viewed as a morbid or clinical episode is not always easy.

The psychiatrists' difficulty in determining the precise point at which depressed mood merges into *clinical depression* is analogous to the internists' dilemma regarding hypertension: at what point can oscillations in blood pressure be considered to herald the onset of hypertensive disease? The psychiatric clinician is usually guided by the following criteria:

1. The *intensity* of the mood change is such that it pervades all aspects of the person and his or her functioning (resulting in role and vocational impairment), and the sufferer can no longer be distracted by simple reassurance.
2. The mood may arise in the absence of any discernible precipitant, or it may be grossly out of proportion to those precipitants that the patient or relatives report as significant.
3. More importantly, the pathological mood appears to persist *autonomously*. As William James, himself a sufferer of the malady, suggested in our opening quotation, the *quality* of the mood change is different from that experienced in normal sadness.
4. Moreover, in morbid affective states, the change in mood is typically accompanied by a cluster of signs and symptoms constituting the affective *syndromes* described in the beginning of this chapter.
5. Finally, in a depressed individual, history of past episodes of elation and hyperactivity strengthens the suspicion that the mood is part of a pathological process.

The mood change associated with a manic psychosis is usually so flagrantly inappropriate to the prevailing life situation of the individual, that family, friends, policemen (who often have to be called to persuade the patient without insight to seek help), and clinicians have no difficulty in diagnosis. A problem may arise in identifying a first episode of hypomania, however. Here the increased activity and the elevated mood appear at first to be a valuable response to interpersonal and professional demands. Nevertheless, sooner or later (unhappily not always before a conjugal or a financial disaster), the precarious and unpredictable actions of the hypomanic individual betray an underlying mood disorder. Sometimes the individual will have difficulty coping with the excessive "nervous" energy, which ceases being pleasant, and this may lead to alcohol or sedative-hypnotic use in an attempt at self-treatment. Even more difficult to diagnose, and often masked by significant interpersonal or vocational problems and substance abuse, are cyclothymic (shorter, milder, and lifelong forms of bipolar illness) and hyperthymic (intermittently hypomanic) conditions. Again, we shall explore these more subtle dimensions during our discussion of classification in Chapter 3.

The Subjective Experience of Depressed Mood

The word "depression" is used variously when describing the experience of mood. An adequate descriptor of the pathological mood must convey both the presence of a painful emotion and the absence of pleasure (anhedonia). It could be argued that these two descriptions refer to the same emotional state, with the pain of depression preventing the experience of pleasure. However, clinical observation does not support this conclusion.[29] Many dysphoric individuals are able to experience pleasure, and not all patients who report the absence of pleasure consider themselves depressed. The painful dimension of depressive experience during illness is usually related to anxiety, guilt, anguish, and restlessness—an agitated state of emotional arousal. Robert Burton,[30] in his book *The Anatomy of Melancholy* (1621), describes the state thus:

> They are in great pain and horror of mind, distraction of soul, restless, full of continual fears, cares, torments, anxieties, they can neither drink, eat, nor sleep for them, take no rest, neither at bed nor yet at board....

Contrast this to Shakespeare's[31] description of Hamlet's depressive anguish:

> How weary, stale, flat and unprofitable
> Seem to me all the uses of this world.

Perhaps the descriptions can best be understood as exemplifying the two poles of a continuum, with the desperate agitated state of arousal described by Burton as a sharp contrast to the anhedonia, apathy, emptiness, and depersonalization gripping and inhibiting the emotions of Hamlet.

It is tempting to speculate that these two forms of depressive experience, the painful and the anhedonic, are temporally related. They are certainly reminiscent of the biphasic response of human and rhesus monkey infants to separation from the nurturing parent (see Chapter 4). The initial response is one of agitated, restless anguish, a behavior that might be interpreted as a seeking of the lost parent. Over time a stage of despair develops, characterized by decreased initiative, loss of interest, withdrawal, and apathy.[32] Both psychoanalytic and behavioral theorists suggest that anxiety is an early component of what individuals identify subjectively as depressive experience. As a first line of defense or reaction to an aversive event, anxiety contains some hope of resolution. However, as the emotion becomes increasingly painful and agonized, the individual despairs of attempts to solve personal reverses. The withdrawn, passive, inhibited, helpless, and dejected state of depression which then supervenes may be seen as a reflection of this resignation of the individual, an adaptation to overwhelming challenge. Beck[8] has argued that this cognitive state of helplessness and hopelessness represents the core of the depressive experience and precedes, in an etiological sense, the pathological syndrome of depressive illness.

In our own experience and practice we have found the following distinction helpful. The painful *mood* termed *depression* is a ubiquitous, somewhat undifferentiated and nonspecific emotional reaction to personal adversity which is seen in a variety of clinical and nonclinical settings. The mood state that is described appears to be subjectively close to that of anxiety. It is an indication of psychic conflict and pain which leads to arousal, restlessness, difficulty falling asleep, and unpleasant dreams. It is seen in the failures of everyday life, bereavement, physical illness, withdrawal from a variety of drugs, and is part of almost all neurotic and psychotic illnesses. In contrast, *anhedonia*—an inability to anticipate and respond to reward—is characteristic of a more restricted range of clinical disorders and is one of the core manifestations of depressive illness.

The Affective Experience: Teleological and Literary Reflections

Transient depressive mood, without the anhedonic component, is indeed a universal human experience. It seems an inevitable accompaniment of everyday stresses and may follow minor illness, such as the common cold or influenza. Temporary dejection, irritability, and anxiety may also be associated with complex psychobiological states such as the premenstrual and postpartum phases. These latter reactions, respectively referred to as premenstrual and maternity "blues," are common enough to be considered as part of normative experience in most cultures. Most transient episodes of depressive mood, however, appear to follow situations involving *separation* from important people, places, or things, from *loss* or perceived loss of special attachments or ideals. Why has this behavioral reaction of pain and withdrawal to loss persisted? What possible evolutionary significance can it have? After all, Darwin's theory tells us that natural selection submerges and destroys characteristics without value to the organism.

It has been postulated that the separation reaction persists as an important biobehavioral defense against the disruption of the essential bond between the nurturing mother and infant. Through the work of Harlow[33] at the University of Wisconsin, for example, we have learned that the distressed behavior of an infant monkey upon temporary separation from a maternal figure elicits nurturing behavior by other adult females, ensuring that the physical and emotional needs of the infant are sustained. Thus depressive withdrawal may be seen as an important form of social communication.

Another interpretation is the "conservation–withdrawal" hypothesis developed by psychiatrists Arthur Schmale and George Engel[34] of the University of Rochester. From their research into infant behaviors, they postulate that the withdrawal following loss conserves the inner resources of the individual to permit subsequent adaptation to new environmental challenges and opportunities.

Whatever teleological explanation is chosen, there is no doubt that disturbances of mood—especially depressive phenomena—are very persistent in all cultures. In our own culture, the experience of sadness has become inextricably interwoven with everyday life and with art and literature.[35]

Some believe that a heightened sense of man's existential dilemma is fostered by the experience of illness and that a special relationship exists between creativity and psychiatric disorder. Recently, Andreasen, a psychiatrist and doctor of philosophy in literature, and her colleague Canter,[36] have found in a detailed investigation of the families of creative writers that two-thirds of the sample suffered from affective illness. A high incidence of morbidity was also seen in the first-degree relatives. Aristotle appears to have agreed when he wrote[35]

> Those who have become eminent in philosophy, poetry, politics and the arts all have tendencies toward melancholia.

In recent times, prominent individuals such as Charles Darwin, Sören Kierke-gaard, Abraham Lincoln, Ernest Hemingway, Virginia Woolf, Winston Churchill, William James, Vivian Leigh, and Sylvia Plath all appear to have struggled with affective disorder.

One may also find some of the best descriptions of mood disorder in literature and art. Robert Burton,[30] about whom we shall hear more in the next chapter, felt it was important to inform his readers that he wrote about melancholia from personal experience:

> [Other authors] get their knowledge by books, I by melancholizing. That which others hear of or read of, I have felt and practiced myself.

Similarly, one finds an awareness in literature of elevated mood as an adaptive response to unpleasant life events. The following passage from Alexander Solzhenit-syn's documentary novel *The Gulag Archipelago* is a clear illustration of such recognition[37]:

> We were marched again through the same magical garden filled with summer—again to the baths! We were overcome with laughter that had us rolling on the floor. This cleansing, releasing laughter was, I think, not even sick but a viable defense and salvation of the organism.

Obviously, as Andreasen cautions, we should not generalize from a famous few that all talent and genius stem from the pathological. However, when man comes to describe himself within the special framework of his subjective relationship to the larger world, science and literature each become part of the total awareness. Literature provides a perspective that can give the student of human behavior valuable insight into the complexity of psychopathology.[38] The historical perspective offers a similar opportunity, one that we shall take up in our next chapter.

References

1. James, W. (1923). *The varieties of religious experience.* Glasgow: William Collins Sons, 1960.
2. Helgason, T. Epidemiological investigations concerning affective disorders. In M. Schou, and E. Stromgren, (Eds.), *Origin, prevention and treatment of affective disorders.* New York: Academic Press, 1979, pp. 241–255.
3. Kraepelin, E. *Manic–depressive insanity and paranoia.* Edinburgh: E. and S. Livingstone, 1921.
4. Lief, A. (Ed.). *The commonsense psychiatry of Dr. Adolph Meyer.* New York: McGraw-Hill, 1948.
5. Lewis, A. Melancholia: A clinical survey of depressive states. *Journal of Mental Science,* 1934, *80,* 277–378.
6. Akiskal, H. S., and Puzantian, V. R. Psychotic forms of depression and mania. *Psychiatric Clinics of North America,* 1979, *2,* 419–439.
7. Himmelhoch, J. M. Mixed states, manic–depressive illness, and the nature of mood. *Psychiatric Clinics of North America,* 1979, *2,* 449–459.
8. Beck, A. T. *Depression: Clinical, experimental and theoretical aspects.* New York: Harper and Row, 1967.
9. Cassano, G. B., Maggini, C., and Akiskal, H. S. Short-term, subchronic, and chronic sequelae of affective disorders. *Psychiatric Clinics of North America,* 1983, *6,* 55–67.
10. Dunner, D. L. Rapid-cycling bipolar manic depressive illness. *Psychiatric Clinics of North America,* 1979, *2,* 461–467.

11. Robins, E. *The final months: A study of the lives of 134 persons who committed suicide.* New York: Oxford University Press, 1981.
12. Murphy, G. E., and Robins, E. Social factors in suicide. *Journal of the American Medical Association,* 1967, *199,* 81–86.
13. Barraclough, B., Bunch, J., Nelson, B., and Sainsbury, P. A hundred cases of suicide: Clinical aspects. *British Journal of Psychiatry,* 1974, *125,* 355–372.
14. Watts, C. A. H. *Depressive disorders in the community.* Bristol: John Wright and Sons, Ltd., 1966.
15. Weissman, M. M., and Myers, J. K. Affective disorders in a U.S. urban community. *Archives of General Psychiatry,* 1978, *35,* 1304–1311.
16. Fenichel, O. *The psychoanalytic theory of neurosis.* New York: W. W. Norton and Co., 1945.
17. Woodruff, R. A., Clayton, P. J., and Guze, S. B. Is everyone depressed? *American Journal of Psychiatry,* 1975, *132,* 627–628.
18. Lehmann, H. E. The impact of the therapeutic revolution on nosology. In P. Doucet, and C. Laurin, (Eds.), *Problematique de la psychose.* New York: Excerpta Medica Foundation, 1969, pp. 136–153.
19. Baldessarini, R. J. Frequency of diagnoses of schizophrenia versus affective disorders from 1944 to 1968. *American Journal of Psychiatry,* 1970, *127,* 759–763.
20. Weissmann, M. M., and Klerman, G. L. Sex differences and the epidemiology of depression. *Archives of General Psychiatry,* 1977, *34,* 98–111.
21. Orley, J., and Wing, J. K. Psychiatric disorders in two African villages. *Archives of General Psychiatry,* 1979, *36,* 513–520.
22. Leighton, D. C., Harding, J. S, Macklin, D. B., Hughes, C. C., and Leighton, A. H. Psychiatric findings of the Stirling County study. *American Journal of Psychiatry,* 1963. *119,* 1021–1032.
23. Hudgens, R. W. *Psychiatric disorders in adolescents.* Baltimore: Williams and Wilkins, 1974.
24. Carlson, G., and Strober, M. Affective disorders in adolescence. *Psychiatric Clinics of North America,* 1979, *2,* 511–526.
25. Post, F. Affective disorders in old age. In E. S. Paykel, (Ed.), *Handbook of affective disorders.* New York: Guilford Press, 1982, pp. 393–402.
26. Hollingshead, A. B., and Redlich, F. C. *Social class and mental illness: A community study.* New York: John Wiley and Sons, 1958.
27. Welner, A., Liss, J. L., and Robins, E. Psychiatric symptoms in white and black inpatients. II. Follow-up study. *Comprehensive Psychiatry,* 1973, *14,* 483–488.
28. Adebimpe, V. R. Overview: White norms and psychiatric diagnosis of black patients. *American Journal of Psychiatry,* 1981, *138,* 279–285.
29. Klein, D. F. Endogenomorphic depression: A conceptual and terminological revision. *Archives of General Psychiatry,* 1974, *31,* 447–454.
30. Burton, R. *The anatomy of melancholy.* New York: Vintage Books, 1977.
31. Rowse, A. L. *The annotated Shakespeare,* Vol. III. *The tragedies and romances.* New York: Potters, 1978.
32. McKinney, W. T., Suomi, S. J., and Harlow, H. F. Depression in primates. *American Journal of Psychiatry,* 1971, *127,* 49–56.
33. Harlow, H. F. Age-mate or peer affectional system. *Advances in the Study of Behavior,* 1969, *2,* 333–383.
34. Schmale, A. H., and Engel, G. L. The role of conservation-withdrawal in depressive reactions. In E. J. Anthony, and T. Benedek, (Eds.), *Depression and human existence.* Boston: Little, Brown, 1975, pp. 183–198.
35. Klibansky, R., Panofsky, E., and Saxl, F. *Saturn and melancholy: Studies in natural philosophy, religion and art.* Nendeln, Liechtenstein: Kraus Reprint, 1979.
36. Andreasen, N. J. C., and Canter, A. The creative writer: Psychiatric symptoms and family history. *Comprehensive Psychiatry,* 1974, *15,* 123–131.
37. Solzhenitsyn, A. I. *The Gulag Archipelago: An experiment in literary investigation.* New York: Harper and Row, 1973.
38. Stone, A. A. and Stone, S. S. *The abnormal personality through literature.* Englewood Cliffs: Prentice-Hall, 1966.

2

Mood Disorders: Historical Perspective and Current Models of Explanation

Such as have the Moon, Saturn, Mercury misaffected in their genitures; such as live in over-cold or over-hot climes; such as are born of melancholy parents; as offend in those six non-natural things, are black, or of a high sanguine complexion; that have little heads, that have a hot heart, moist brain, hot liver and cold stomach, have been long sick; such as are solitary by nature, great students, given to much contemplation, lead a life out of action, are most subject to melancholy. Of sexes both, but men more often; yet women misaffected are far more violent, and grievously troubled. Of seasons of the year, the autumn is most melancholy. Of peculiar times: old age, from which natural melancholy is almost an inseparable accident; but this artificial malady is more frequent in such as are of a middle age. Some assign forty years, Gariopontus thirty. Jobertus excepts neither young nor old from this adventitious.

Robert Burton[1]
The Anatomy of Melancholy (1621)

Robert Burton was a parson. Coming as a student to Braenose College in Oxford in 1593, he was elected to Christ Church in 1597 and spent the rest of his life there, most of it in his study. Holbrook Jackson in the Introduction to the 1932 edition of the *Anatomy*[1] describes Burton as a "good humored pessimist—a recluse but no hermit," an example of how "diligence and an enjoyment of drudgery can accomplish miracles in the spare time of a busy life." Sir William Osler considered the *Anatomy of Melancholy* as the greatest medical treatise ever written by a layman. However, it is now found in most bookstores on the shelf marked "English Literature," looking in today's paperbacked world somewhat too large for its cover. Its many pages explore depression conceived in the broadest sense, from "love melancholy" to the "causeless sadness" of the psychotic depressive.

In the early 17th century "melancholy" was the general notation for a broad range of ills that plagued mental life. Nonetheless, we find uncanny similarity in Burton's work with current thinking. Having first excluded hydrophobia (rabies), wolf-madness, phrenitis (where the mind is "annexed" by an acute fever), St. Vitus' dance, and other such common ailments, he went on to divide melancholy into two

Table 2.1 Concepts of the Mind and Depression prior to the Early 1900s

Major proponent (period)	Model	Place in history of ideas	Clinical implications
Hippocrates (fourth century B.C.)	Melancholia	Reductionistic, naturalistic	Mental disorders are due to natural causes
Plato (fourth century B.C.)	Divine madness	Supernatural	Madness being divine inspiration, it is a creative force
St. Paul (first century A.D.)	Rational vs. divine depression	Free will vs. determinism	Insight leading to confession and relief
Johann Sprenger and Heinrich Kraemer[a] (15th century)	Diabolic possession	Supernatural	The mad are evil rather than ill
René Descartes (17th century)	Cartesian	Dualistic, reductionistic	Bifurcation of psychological and biological approaches to man
Robert Burton (17th century)	Habit vs. disposition	Pluralistic, empirical	Multiple factors enter into the causation and treatment of melancholia
Philippe Pinel (18th century)	"Moral"[b]	Humanistic, empirical	Psychological suffering is a respectable subject of medical study
J. E. D. Esquirol (19th century)	"Moral"	Clinical science, empirical	Establishment of sad melancholy as a primary disorder of mood
Emil Kraepelin (19th/20th century)	Disease	Descriptive, biological	Natural history in the differentiation of major psychiatric disorders
Sigmund Freud (19th/20th century)	Metapsychology	Reductionistic, dynamic	Interaction of drive and experience, importance of early experience and drives
Adolf Meyer (19th/20th century)	Psychobiology	Holistic (with psychosocial emphasis)	The birth of social casework

[a]The two Dominican clergymen who wrote *Malleus Maleficarum*, the notorious handbook on witch-hunting, utilized by the Inquisition during the latter part of the medieval era.
[b]"Moral," until the latter part of the 19th century, was synonymous with "psychological" as used in contemporary psychiatry.

major categories—habit or disposition and the nature of being mortal. We would probably today equate the latter with the existential anguish of everyday life. The melancholy of "habit" described in the opening quotation of this chapter, however, bears a striking resemblance to the severe depression we outlined in Chapter 1.

In this chapter we shall principally concern ourselves with the conception of affective illness dominant over the past two centuries, from approximately the early 1800s to the present day. However, earlier concepts of mental illness and mental life are essential to a comprehensive understanding of the current approaches to affective illness. Hence we must spend a little time reviewing those also. Table 2.1 will serve as a useful catalogue in this regard.

Early Struggles: The Empirical and the Supernatural

The cold black disposition of the melancholic, the bilious origin of the disorder—ideas that are still in our language today and that were dominant in Burton's time—have a long history. Indeed, at times the concept of the elemental origin of melancholia seemed almost immortal, guiding medical practice for centuries. The four-element theory, the basis of the physiological doctrines of the Hippocratic school, was initially formulated by Empedocles (490–430 B.C.), a Sicilian by adoption. The four basic elements of the earth: fire, water, air, and the earth itself were thought to be reflected in the human body as the basic qualities of heat (blood), dryness (phlegm), moisture (yellow bile), and cold (black bile), and were believed to be found in the heart, brain, liver, and spleen respectively. It was the balance of these humors that broadly determined health and illness.

Supernatural influences, both divine and mythological, also played a dominant role in the explanation of mood disorders, as well as other forms of mental illness. Melancholics were believed to be born under the sign of Saturn[2] and dominated by the black bile secreted by the spleen; this organ, it was postulated, occupied the same position in the human organism as the planet Saturn in the firmament.

It was generally accepted that divine will played the major role in determining those who would suffer insanity. In Deuteronomy, Chapter 28, verse 28 we read that "the Lord shall smite thee with madness and blindness and astonishment of heart."

Similar concepts are found in the epic poetry of Homer: At the battles of Troy Ajax fights valiantly and scorns the help of the gods, particularly Athene, the protectress of Diomedes and Odysseus. When the armor of the dead Achilles is given to Odysseus, Ajax becomes angry; Athene deprives Ajax of his faculties, torturing him with delusions which drive him to slaughter cattle and sheep in the belief that they are his enemies. On regaining his senses, Ajax is so overcome with feelings of remorse and guilt that he throws himself on his sword and dies a suicide.[3]

In those early days, insanity was not always considered a detriment, however. When Plato (427–347 B.C.) wrote in *Timaeus,* his dialogue of physics and biology, that prophetic truth seldom occurred to a man in his right mind, he was concurring with a particular view of "divine madness," that inspiration to genius and prophecy could be found in the human madness instilled by God.[4]

It is against such a background that the work of the Hippocratic school should be placed. At the time, in the fourth century before Christ, it came as a breakthrough for empiricism. Hippocrates (460–355 B.C.) argued that mental disorders and epilepsy—the latter known as the "sacred disease" and long considered the prototype of divinely inspired madness—were due to brain dysfunction:

> And men ought to know that from nothing else but thence (the brain) come joys,
> delight, laughter and sports, and sorrows, griefs, despondence, and lamentations. . . .

> And by the same organ, we become mad and delirious, and fears and terrors assail us, some by night, and some by day, and dreams and untimely wanderings, and cares that are not suitable and ignorance of present circumstances. . . . All these things we endure from the brain when it is not healthy.[5]

The history of psychiatry from Hippocrates to the French Revolution—which heralds the modern era of psychiatry—is that of a pendulum swinging between these two poles: the supernatural and the empirical. During the medieval period, however, empiricism was to be virtually submerged, and by almost universal consensus metaphysics held sway.

Thus when Albrecht Dürer was born in 1471, the philosophic pendulum had drawn close to the metaphysical pole. Dürer was the second and favorite son of a poor, brooding Nuremberg goldsmith. The son was to grow in his father's likeness, tormented by genius and despair. Rapidly successful as a painter, Dürer turned to line engraving in wood and copper as he grew older. It was a century before the empircism of such thinkers as Harvey, Descartes, and Burton, and Europe was firmly locked in the twilight struggles of the Middle Ages. The Roman Church had mounted the Inquisition as a necessary evil to destroy an even greater threat, that of paganism and an ensuing apocalypse. Dürer had been in adolescence when *The Malleus Maleficarum*[6] was produced to support the papal bull of Pope Innocent VIII (1484). It included the following notation:

> It has indeed come lately to Our ears, not without afflicting Us with bitter sorrow that in some parts of Northern Germany, as well as in the provinces, townships, territories, districts and dioceses of Mainz, Cologne, Tréves, Salzburg, and Bremen, many persons of both sexes unmindful of their own salvation and straying from the Catholic Faith, have abandoned themselves to devils, incubi and succubi. . .

Madness was still considered supernatural, but was entirely evil. Thousands of innocent men and women, normal and deranged, were tortured and killed in what was seen as the exorcism of a devil's army intent upon human perversion. The situation was to worsen when in 1499 the respected astronomer Lohomnes Stoffler predicted a planetary conjunction in the sign of Pisces which in February 1524 would produce a deluge even more destructive than the great biblical flood. It was considered by many as a sign that the end of the world was near. As the time approached, those who were able to do so began to build boats and shelters on high ground, swindlers convinced owners to leave their property and then plundered the contents, pamphlets were written to advise and to deny. It was in this climate, at the age of 43, that Dürer first printed his engraving of Melancholia I (see Frontispiece).

The engraving is Dürer's conflicted call for trust in God and faith in the penitence of mankind. It is also a graphic window on the concept of melancholia in the early 16th century. Melancholia is depicted in two forms.[7] The first is the winged bat, an allegory of the forces of evil, flying from the picture as light from a rainbow (a biblical sign of divine providence) and a passing comet illuminate the body of water below. The crouched human figure—surrounded by the tools of the builder and sitting before the unfinished tower—represents the despair of earthly pride and genius. The figure, resolute but preoccupied with a fixed stare, has withdrawn from the commerce of everyday life. It is this element of the 400-year-old etching that particularly catches the essence of severe depression as we still see it today. Apparently Dürer himself was afflicted and believed that through the atonement of sin, so well known (and frequently only known) to the melancholic, human effort could be turned from self-aggrandizement to the work of God. Evil would be then driven from the world.

In those discordant times we thus find a cultural consensus. Affective illness was a consequence of man's daily thought, an atonement for the sin of pride. Man was an extension of God and his church, and the explanation of mood disorders thus came logically from an understanding of that being. There was little need for further explanation.

Introspection and the Comfort of Metaphysics

How we structure what we know about ourselves as human beings will always be, in part, a function of history. Patterns of understanding emerge slowly and, although they may be conceived by individuals, they become accepted and powerful only through cultural consensus. Hypotheses are generated, assailed, and modified through succeeding generations. Some are suppressed, others just simply lost, to emerge apparently *de novo* decades or centuries later.

Thus, in context, one must have some sympathy for our medieval forebears. Human beings *are,* after all, both the most wondrous *and* the most proud of living creatures. In the search for our subjective selves, we still have little doubt that we are above the animal herd, unique in the animal kingdom. Further, with our capacity for symbolic abstract thought and associated language, we probably do have a unique intellectual awareness of our transience in this material world. In the Middle Ages, with only a primitive knowledge of biology, it is hardly surprising that man saw God and himself as indivisible. It was to most a very comforting viewpoint, providing both meaning and a sense of order to human existence at a time of enormous social upheaval.

The Cartesian Dualism of Mind and Body

Even at the beginning of the 17th century, the prevailing dictum remained that the structure of the universe and man defied analysis. Since exact *verbal* description was impossible, the universe was considered imponderable and thus little worthy of study.

René Descartes, like other thinkers of the period, was the product of a Jesuit education. He had, however, developed an unusual fascination with mathematics, particularly with geometry. As the story goes, when so preoccupied, upon a November night in 1619, it occurred to him that if the structure of the universe were logical, then it must be open to abstract understanding through mathematical analysis. Descartes was then 23 years old.

Later, as the preface to a book on the geometry of the planets, he outlined a "discourse on [his] method."[8] This essentially offered four rules of basic logic with which to unmask the imponderable questions of Aristotle: to maintain a posture of doubt, to divide every problem into as many parts as possible, and only then, by reflection upon these elements, to commence the analytic task. Descartes, of course, was describing a method that has come to be known as reductionism, the cornerstone of our modern scientific investigation.

While Descartes' ideas were formulated during a time of rekindled interest in the workings of nature and of the universe, new ideas were still not easily accepted. The *Malleus Maleficarum,*[6] that awesome guide to witch-hunting, had been selling well since 1486 and went through 16 revisions and editions during Descartes' lifetime.

The quest of science was of particular concern to the established church. Descartes' new methods of inquiry cast an uncomfortable light on the special relationship of man to God. Increasing knowledge of the material world had begun to raise doubts regarding the origins and destinies of man. Descartes, perhaps in conscious recognition of this dilemma, developed an ingenious way out. Mind—the soul of man—was not to be considered dependent for survival upon the earthly body. He postulated two parallel but independent worlds, one of mind and one of matter, each of which could be studied without reference to the other. The mind and the body were analogous to autonomous clocks which were nonetheless coordinate and kept the same time. Although Descartes' clever solution was considered indefensible even by some of his contemporary philosophers in separating psychological functioning so completely from physical functioning, it served to foster the nascent discipline of investigation. Freed from theological interference, empirical science began to make considerable progress. A rigid preconception of the universe and world gave way to a meaningful and objective understanding of its elements, including the nature of man.

The Rise of Humanism and Objective Observation

Descartes' dualism served to allow the links between mental derangement and the devil to be broken. Slowly the chain was reforged to again connect madness and brain. This cleared the way for a reconsideration of the insane as human and therefore in need of "moral" (i.e., humane psychological) rather than exclusively spiritual treatment. The French physician Philippe Pinel, famed for having unchained the mentally ill, based his humanitarian reforms on the practical guidance provided by Poussin, an ex-patient risen to the rank of Superintendent at the Salpêtrière Hospital. However, in his humanism Pinel was clear also in his intent to apply the Cartesian method of observation.

> From the systems of nosology I have little assistance to expect; since (they are) better calculated to impress the conviction of their insufficiency than to simplify my labor. I therefore, resolve to adopt that method which has invariably succeeded in all the departments of natural history, viz. to notice successively every fact, without any other object than that of collecting materials for future use; and to endeavor as far as possible to divest myself of the influence, both of my own prepossessions and the authority of others.[9]

Hence, in the early 19th century, the dualism of mind and body, which we now correctly consider a stumbling block for the integration of ideas, was actually fostering two very important psychiatric advances: a return to humanism and the laying down of the foundation for an objective phenomenology.

France was the crucible for the nascent discipline of psychiatry. Esquirol, Pinel's successor at the Salpêtrière, was one of the most talented and systematic observers of the time. He was among the first to suggest that sad melancholy was a primary disorder of the "passions" (i.e., mood), rather than being secondary to "insanity," or impaired reason.[10] Later Falret described a cluster of psychotic disorders that were periodic, or *"folie circulaire."* Baillarget, with even greater vision, proposed the alternating forms of mania and melancholia as *"folie à double forme."* These physicians were the forerunners of Emil Kraepelin, a German (1855–1926), who later in the century was to vigorously classify mental disorders by studying the clinical course and outcome of the various symptom complexes.[11]

The symptoms of mania and depression at first seem antithetical. However, examination of their clustering together, the lack of mental deterioration in periods between the episodes, and the essential periodic nature of both syndromes led Kraepelin to confidently divide "manic–depressive insanity" or disorders of mood from "dementia praecox" (later to be called schizophrenia by Bleuler). Clinical pictures that had previously been called alternating insanity, cyclothymia, periodic depression, periodic mania, simple mania, and simple melancholia were all now clustered under this new term. Later, under scientific pressure from his students, Kraepelin added "involutional melancholia" to this list.

The Progenitors of Current Understanding: Freud, Kraepelin, and Meyer

Eighteen ninety-six was an important year from the standpoint of our historical review. *Emil Kraepelin* outlined his classification of mood disorders in the fifth edition of his successful *Lehrbuch*. At the age of 40 he was a respected professor teaching and conducting research in Heidelberg. *Sigmund Freud* was also 40 years old, and working in Vienna. His early career in neurology and his collaboration with Breuer were terminating as he sought ways other than hypnosis to research the human psyche. *Adolf Meyer* (1866–1950) had left Switzerland for the United States and had just been appointed at the Worcester State Hospital in Massachusetts. There he was immersed in the study of neuropathology. These three men, each a progenitor of our current concepts in psychopathology, were already following their divergent paths.

During the 19th century, concepts of mental illness had in fact reached a startling conformity. The nature of this consensus is best understood against a background of impressive progress that had occurred in general medicine. It had been the century of Pasteur, of Lister, and of Virchow, of cleaner water supplies, sewage systems, microscopes, an understanding of microorganisms, and morbid anatomy. Overall the structural approach to disease, that of organic pathology, had paid handsome dividends in improving public health. Careful diagnosis and the study of natural history had given the physician the power of prediction. Furthermore, when tragedy struck, the physician's accuracy and ability could be verified at postmortem. With the brain again recognized as the seat of madness, there was reason to believe that neuropathology could do for psychiatry what general pathology had provided for the rest of medicine. Along with their contemporaries, Freud, Kraepelin, and Meyer all initially shared these vital hopes and aspirations. However, while by the end of the century the spirochete of syphilis could be clearly held responsible for the insanity of general paresis, a clear basis in neuropathology for most mental dysfunction, including the affective disorders, remained elusive.

Kraepelin did not lose hope. He continued to refine his diagnostic categories, remaining convinced that mental disturbance, just as any other human ailment, must have within it categories of illness with varying neuropathology (or, as we might say today, pathophysiology). For him, elemental observation and the study of outcome held the potential power of prediction.

Freud and Meyer, however, were disillusioned and sought other dimensions that promised more immediate understanding. Instead of looking forward in time, they both began to look back, asking such questions as "How did this individual become this way? What in the biography of the individual might explain the present behaviors?" It was an attempt to explain present illness by antecedent psychopathology rather than waiting for the future to validate a diagnosis. By contrast, the school from which Kraepelin had emerged as the leader accepted the occurrence of "psychogenic" states of depression occasioned by situational misfortune, but laid greater emphasis on hereditary concepts of etiology. The work of the Kraepelinians, in fact, embodied without question the Cartesian dualism of mind and body, and paid little heed to emerging physiological concepts such as homeostasis, adaptation, and defense. Disorders of mood, for Kraepelin, were divided into those psychologically occasioned and those of somatic origin.

Both Freud and Meyer, on the other hand, sought a dynamic integration of what they believed to be a spurious division. Such a task was difficult. The limits of understanding of the brain and nervous system that troubled Kraepelin were potentially an even greater burden to these men. How does one build a dynamic view of mental life when only the crude anatomic structure of brain is understood? Apparently, Freud truly longed for the day when biological understanding would verify his formulations; however, in the absence of such information he felt compelled to develop a metastructure and metaphysiology within which he could explore and explain his emerging thesis of mental life.

The Common-Sense Psychobiology of Adolf Meyer

Adolf Meyer took a middle road between the empiricism of Kraepelin and the metapsychology of Freud. Perhaps as a result, his ideas are less well-known than those of his two contemporaries.

In 1915, in a paper read before the American Medical Association meeting in San Francisco, Meyer outlined his ideas thus:

> Psychobiology forms clearly and simply the missing chapter of ordinary physiology and pathology, the chapter dealing with the functions of the total person and not merely of detachable parts. It is a topic representing a special level of biological integration, a new level of simple units having in common the fact of blending in consciousness, integrating our organism into simple or complex adaptive and constructive reactions of overt and implicit behavior.[12]

In his description and understanding of psychopathology, Meyer sought to strike a balance between reductionism and integration. Also, as a clinician his major goal was to close the gap between our understanding of intrapsychic experience and the sociological, interpersonal, and biological determinants of behavior. His was a crusade against that all too natural preference we have for dividing the metaphysical mind from its biological embodiment.

Adolf Meyer, born in 1866 in a small Swiss village, was no stranger to philosophy or metaphysics. His father, who was a minister, had a respectable library in science and philosophy. When the time came for Adolf to choose a career, he was drawn toward medicine as a middle ground between the sciences and theology. While a schoolboy he had become very interested in the value of the biographical sketch as a way of describing and understanding an individual. This was to be reflected in his emphasis upon the teaching of medical history-taking and the development of his "life chart" (Fig. 2.1).

As a medical student, he was much impressed by Professor Forrell at Zurich, whose interests simultaneously embraced the behavior of ants and brain anatomy! Intrigued by the professor's efforts to integrate such seemingly diverse phenomena, Meyer at the age of 24 spent a year with Hughlings Jackson in London. At the same time he became acquainted with the work of Thomas Huxley. He was a great admirer of these men, dynamic scientists in the tradition of the great French physician and physiologist Claude Bernard. Throughout his career he attempted to emulate them, intertwining objective elemental description with efforts to develop a unifying understanding. In his teaching he was faithful to his goal of meticulous observation as the cornerstone of the understanding of psychopathology. As aids to his students, he developed various mechanisms, including that of the life chart, in which he sought to integrate biological and psychosocial factors over time. He defined health broadly as the reactive and constructive adaptation of a completely integrated organism. Indeed, it is the dynamic paradigm of adaptation over time that distinguishes Meyer's approach to psychopathology from that of his contemporaries such as Kraepelin, Wernicke, and Bleuler. He emphasized that the cardinal feature of the living organism was its reactivity, its adaptation, and that this in itself was worthy of study. We must ask, he said,

> . . . what are the individual's assets, the reactive and associative resources in the form of affective and expressive activity. Under what conditions are they apt to go wrong, and under what conditions can they be modified again for the better. We are dealing with absolutely objective and positive facts peculiar only in the way in which they hang together by association in the wealth of equivalents and combinations and in the varying extent and depth to which they implicate the parts of the integrated organism.[12]

Meyer spent virtually the whole of his professional life in the United States— initially in Illinois, then later at Worcester, Massachusetts, the New York State Department of Hygiene, and finally at Johns Hopkins, where he became the first professor of psychiatry in 1909. In fact, while in Massachusetts he paved the way for the "five lectures on psychoanalysis" which Freud delivered at Clark University in 1909. Meyer, while working at the Worcester State Hospital, had been on the faculty at Clark for almost a decade, and together with Jung and Freud he received a Doctor of Law degree at the time of the Clark Decennial celebration. Meyer chose his work on dementia praecox for presentation, and it is said that after the discussion, he and Freud

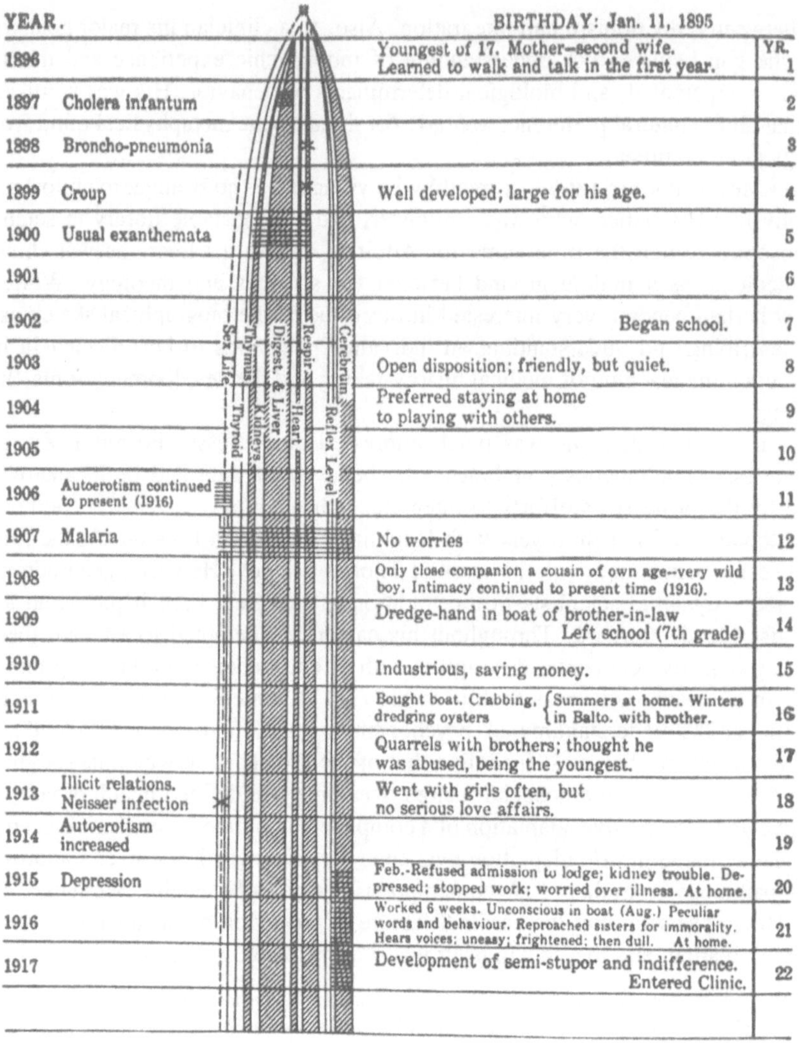

YEAR.		BIRTHDAY: Jan. 11, 1895	YR.
1896		Youngest of 17. Mother—second wife. Learned to walk and talk in the first year.	1
1897	Cholera infantum		2
1898	Broncho-pneumonia		3
1899	Croup	Well developed; large for his age.	4
1900	Usual exanthemata		5
1901			6
1902		Began school.	7
1903		Open disposition; friendly, but quiet.	8
1904		Preferred staying at home to playing with others.	9
1905			10
1906	Autoerotism continued to present (1916)		11
1907	Malaria	No worries	12
1908		Only close companion a cousin of own age--very wild boy. Intimacy continued to present time (1916).	13
1909		Dredge-hand in boat of brother-in-law Left school (7th grade)	14
1910		Industrious, saving money.	15
1911		Bought boat. Crabbing, {Summers at home. Winters dredging oysters {in Balto. with brother.	16
1912		Quarrels with brothers; thought he was abused, being the youngest.	17
1913	Illicit relations. Neisser infection	Went with girls often, but no serious love affairs.	18
1914	Autoerotism increased		19
1915	Depression	Feb.-Refused admission to lodge; kidney trouble. Depressed; stopped work; worried over illness. At home.	20
1916		Worked 6 weeks. Unconscious in boat (Aug.) Peculiar words and behaviour. Reproached sisters for immorality. Hears voices; uneasy; frightened; then dull. At home.	21
1917		Development of semi-stupor and indifference. Entered Clinic.	22

Center chart vertical labels: Sex Life, Thyroid, Thymus, Digest. & Liver, Kidneys, Heart, Respir., Cerebrum, Reflex Level

A Case of Schizophrenia

Figure 2.1 An example of a life chart used by Adolf Meyer in teaching psychobiology: Diagram "A Case of Schizophrenia" from E. Winters (Ed.). *The collected works of Adolf Meyer*, vol. III. Baltimore: Johns Hopkins Press, 1951, p. 55.

exchanged ideas for half an hour or so. The brief encounter culminated with Freud offering Meyer a pamphlet on the id! It was to be their only meeting.

Meyer was initially sympathetic, perhaps mildly enthusiastic, about Freud's theories. In 1906 he wrote, "Freud has opened the eyes of the physician to an extension of human biology which differs very favorably from the sensational curiosity shop of the literature on perversions" (presumably an oblique reference to Krafft-Ebing!). He even went so far as to comment later that Freud's work could become as important in psychopathology as the study of dietetics was to the general practitioner.

Slowly, however, he became disillusioned. He became convinced that Freud's method encouraged "mysticism"—an "esoteric" system which fostered psychiatric denominationalism that required, as he put it, "the instincts of a prosecuting attorney and the talent of a constructive imagination."

Meyer preferred a broader approach to the subject of psychopathology which would ultimately include a dynamic concept of biology as well as mental life. "All biological function is adaptation." The pathological person was one who used ill-adapted makeshifts that undermined the development and maintenance of health. "What is the faulty reaction? What are the conditions that led to it? How does it respond to special tests and attempts at modification? What can we expect to achieve and what are the steps to be taken?" Freud "too largely emphasized a portion of the situation," binding the dynamic concepts only to his structural theories of psychology and avoiding the concomitant biology.[13]

Despite their differences, Meyer and Freud held much common ground in their approach to psychopathology. Together they both sought to break the entrenched belief that hereditary factors and biological degeneration played the major role in the development of mental illness. Maudsley, for example, who earlier had been a crusader in the opposite direction, "to treat mental phenomena from a physiological rather than a metaphysical point of view," by 1895 saw mental illness only as the inexorable unfolding of genetic factors. Meyer and Freud both believed that everyday events impinged upon the individual, forcing adaptation. For Freud the daily assault was frequently placed in conflict with the instinctive drive and the moral sense gained from one's parents; the result was a battleground ripe for the development of psychopathology. Thus Freud's approach was also a dynamic one, recognizing the flux of all living mechanisms, but his major focus was upon the *psychological* mechanisms of defense and adaptation. Meyer chose a more difficult task—that of integrating, into a dynamic whole, *both biological and psychological mechanisms*. He chose, in fact, to walk the knife's edge between the traditionally divided conceptual frames of mind and body—to side with neither Freud and mental mechanisms nor Kraepelin and somatic processes.

Efforts at integration do not easily find integrity. The clarity and complete nature of Freud's thesis doubtless underscored its success. Furthermore, it could be utilized and discussed without waiting for advances in biological technology. Kraepelin's views, well suited to the descriptive traditions of empirical medical science, found continued support, especially in Europe. However, Meyer's ideas, after his death, withered back. He is perhaps now best known as the father of social casework.

Current Concepts of Mood Disorder

We shall now review some of the prevailing "models" of psychopathology as they apply to mood disorders. Not surprisingly, they fall predominantly into groups that have their roots in the conceptual divisions we have been discussing. Furthermore, as in the past, the models appear to compete with each other—each seeks primacy in explaining cause, rather than just describing the phenomenon, of mood disorder from a particular perspective. Hence, the models rarely bridge difficult conceptual areas—in fact, they frequently divide them. Depression has been the primary focus of the majority of the models proposed; far fewer have attempted to account for the phenomena of mania (Table 2.2).

In common with other psychopathological phenomena, ideas about mood disorders can be placed within frames of reference that represent the intrapsychic, behavioral, and biological perspectives. The *intrapsychic perspective*—as exemplified by the psychoanalytical, existential, and cognitive schools—traces the index experience of depression to past, often remote, events as they impinge on the emotional, cognitive, and experiential world of the individual. The *behavioral perspective,* by contrast, focuses upon the depressionogenic impact of interpersonal or other social events external to the individual. Both perspectives, the intrapsychic and behavioral, utilize purely psychological constructs to explain the origin of depression. The biological perspective, on the other hand, predominantly defines the physiochemical events thought to underlie or predispose to mood disturbances. Thus in most models the mind–body dichotomy is perpetuated. Only rarely is integration attempted.

Intrapsychic Models of Depression

The unique contribution of the intrapsychic perspective is its emphasis upon the significance of personal experience in the development of depressive phenomena.[14] Also important is the dynamic nature of most of the models, where adaptation and psychological mechanisms of defense are perceived as key concepts.

Depression as Retroflexed Anger

The view that depression is caused by aggression turned inward or against the self is one of the most widely quoted psychological interpretations of depression. It represents one of the few psychoanalytic hypotheses that did not originate with Sigmund Freud, although he was intimately involved in its ultimate formulation. It was one of his Berlin disciples, Karl Abraham, who first enunciated the hypothesis that depression results from the inward turning of aggression (cf. Chapter 5).[15] The retroflexed anger is actually directed against the loved person who has thwarted the patient's need for dependency and love. Because the patient has internalized the loved object in his attempt to prevent the traumatic loss, he himself unwittingly becomes the target of the destructive anger. In the extreme, the process may end in self-destructive or suicidal behavior. Again, the punishment is intended for the internalized object that is both loved and hated in an ambivalent fashion.

From a historical perspective, the crucial concept here is the derivation of one affect—depression, in this instance—from another, i.e., anger, with the conversion taking place in the closed space of the mind. In Freud's earlier work, the origin of anxiety had been similarly conceptualized by the transformation of unexpressed, dammed-up sexual energy. Although a biological force (sexual and aggressive energy) is involved in these energy transductions, the model of a hydraulic mind is really a metaphor. Freud did not infer a real physiochemical space in the anatomy of the brain,

Table 2.2 Contemporary Models of Depression

Proponent (year) (Ref. no.)	Model	Mechanism	Scientific and clinical implications
Karl Abraham (1911) (15)	Aggression-turned-in-ward	Transduction of aggressive instinct into depressive affect	Hydraulic mind, closed to external influences (Non-testable)
Sigmund Freud (1917) (17) John Bowlby (1960) (18)	Object loss	Disruption of an attachment bond	Ego-psychological Open system Testable
Edward Bibring (1953) (19)	Loss of self-esteem	Helplessness in attaining goals of ego ideal	Ego-psychological Open system Social and cultural ramifications
Aaron Beck (1967) (21)	Cognitive	Negative cognitive schemata translate into depressed mood and behavior	Ego-psychological Open system Reductionistic Testable Predicts phenomenology Suggests treatment
Martin Seligman (1975) (24)	Learned helplessness	The belief that one's responses will not bring relief from undesirable events	Testable Predicts phenomenology Predicts treatment
Peter Lewinsohn (1975) (27)	Loss of reinforcement	Low rate of reinforcement, specifically reinforcement presented noncontingently; or social deficits that preclude responding to potentially rewarding events	Testable Predicts phenomenology Predicts treatment
Arthur Prange (1964) (28) Joseph Schildkraut (1965) (30) William Bunney and John Davis (1965) (29) Alex Coppen (1968) (31) I. P. Lapin and G. F. Oxenkrug (1969) (32) David Janowsky et al. (1972) (33)	Biogenic amine or neurotransmitter imbalance	Impaired monoaminergic transmission or cholinergic dominance	Testable Reductionistic Explains phenomenology Suggests treatment
Alex Coppen and D. M. Shaw (1963) (36) Peter Whybrow and Joseph Mendels (1968) (38)	Neurophysiological	Hyperarousal secondary to intraneuronal sodium leakage	Testable Reductionistic Explains phenomenology Suggests treatment
Hagop Akiskal and William McKinney (1973) (39) Peter Whybrow and Anselm Parlatore (1973) (40)	Final common pathway hypothesis	Stress-biology interaction converging on reversible deficits in the neurophysiological substrates of reward	Testable Integrative Pluralistic Explains phenomenology Suggests treatment

although he hoped that such anatomical localization of psychopathological phenomena would one day be demonstrated.

This concept of the mind implies that conflict arises from the battle of incompatible intrapsychic forces confined in a psyche that is exclusively formed by past experience. There is minimal consideration given to the influence of current forces outside the individual.

While the sexual-energy-transduction hypothesis of anxiety has not prevailed in modern psychoanalytic thought, the aggression-turned-inward model of depression does survive.[16] One inference from such a thesis is that depressed patients would be characterized by a relative lack of assertion and outwardly directed aggression. This aspect of the model appears to be supported by clinical phenomena. However, "hostile" feelings and actions are so variable in depressed persons that one must doubt the value of the thesis as a general explanation of depression, despite its popularity.

Recent developments in psychoanalytic theory place greater emphasis upon the role of the ego in mental function and adaptation. Since the ego is the conceptual interface between the self and the world, the ego-psychological perspective makes considerable allowance for active interchange between the mind and the environment. Thus, intrapsychic formulations of depression have given increasing weight to the importance of actual and symbolic loss and injury to the ego during the individual's complex interaction with conjugal, familial, societal, and cultural forces.

Depression as a Response to Object Loss

In psychoanalytic language, object loss generally refers to traumatic separations from significant objects of attachment. Since the meaning of a loss is a matter of subjective interpretation, it is frequently stated that those losses which appear to be trivial may actually precipitate profound depressive reactions. It is therefore desirable, for the sake of conceptual clarity, to specify which kind of object loss is involved, e.g., separation, loss of a limb, disillusion with one's ideals. In the psychoanalytic literature, love loss and bereavement have traditionally received preferential emphasis.

"Love melancholy" had been described much before Burton. Avicenna, the Persian physician whose thinking dominated the Arab world in medieval times, had already declared it to be a "disease." Romeo and Juliet committed suicide because of it, and Shakespeare's sonnets are devoted to it. It was in Freud's 1917 paper on *Mourning and Melancholia,* however, that the similarities and differences between grief or love melancholy and melancholia were first examined:

> In mourning it is the world which has become poor and empty; in melancholia it is the ego itself.[17]

The object-loss model finds its most extensive and complete exposition in the work of Bowlby in London, England.[18] The attachment bond that the growing human being establishes with the mother or other emotionally nurturing individuals represents the focal point of his thesis. Bowlby argues that this early bond serves as a prototype for all subsequent bonds with other people and with the world at large. Furthermore, the adult predisposition to depression, broadly defined, is held to be directly related to whether early separation from important loved ones was successfully managed. Adult losses, of which separations are the most frequent and potent, are postulated to revive the childhood loss and thereby lead to various forms of psychopathology, including depression. This two-step hypothesis has been subjected to numerous studies which we shall review in Chapter 4.

Conceptually this model, placing the interaction of subject and object in a complex network of erotic, social, and ideologic bonds, emphasizes adaptation and the development of autonomy, as the central task of the growing infant. It is argued that the human organism's preferred adaptive state is that of a secure attachment to "significant others." Hence, the close kin of a bereaved individual, in providing optimal social support, can prevent the progression of grief to clinical depression. Likewise, immigrants to a new world tend to cluster together in search for mutual social support. Such attachments may provide a general resistance to psychobiological decompensation, and the disruption of such bonds is relevant to a broad range of psychopathological disturbances. In Chapter 4 and later in Chapter 9, we shall discuss under which circumstances and in which kinds of individuals such separations carry an increased risk for clinical depression.

In summary, we may conclude that the object-loss model represents a major conceptual advance over its predecessor, the aggression-turned-inward model. In postulating an open interactional system between the organism and its social environment, the model easily accommodates multiple antecedent and concurrent variables that change the depressionogenic impact of the separation itself. Such variables include genetic influences, character structure, and concomitant life events as well as the nature of the supporting environment that remains after the separation occurs.

Depression as Loss of Self-Esteem

Edward Bibring, a New York psychoanalyst, has reformulated the classical psychodynamics of depression, preferring to regard it as an independent ego state unrelated to the vicissitudes of the aggressive drive. Helplessness is the focal point in his formulation; depression supervenes when the ego is cognizant of a goal and simultaneously aware of its inability to attain that goal.[19] Narcissistic wounding of the ego results in a collapse in self-esteem. Depression thus emerges when one cannot live up to one's ego ideals: the wish to be worthy and to be loved, the striving to be secure, strong, and superior, the aspiration to be good, kind, loving, and humane.

To define depression in this way is to define it as a psychosocial phenomenon. The concept of the ego, unlike that of the id, is rooted in social reality, and the ego ideal is composed of socially learned symbols and motives. A breakdown of self-esteem may involve, in addition to object loss, man's symbolic possessions, such as power, status, social role, identity, values, and existential purpose. Depression, therefore, falls particularly upon the overambitious, the conventional, the individual with upward social mobility, and the woman who strongly identifies with a passive social role.[20] Even more than the object-loss model, Bibring's conceptualization provides broad links with man's existential, sociological, and cultural worlds.

Depression as Negative Cognition

The cognitive model[21] develops the full implications of the ego-psychological approach to understanding depression. Aaron Beck, the originator of this approach, is a University of Pennsylvania analyst with broad interests in the psychopathology of depression. Like Bibring, he postulated that helplessness and hopelessness represent the core experiences of clinically depressed individuals. Furthermore, he hypothesized that they cause the emotional state of depression. Thus, according to Beck, the thinking patterns of the depressive are characterized by a peculiar "cognitive triad" of negative conception of the self, negative interpretations of one's experiences, and a negative view of the future. The depressed individual finds the world presenting insuperable obstacles and has given up any hope of exercising future control over his destiny. In its most recent formulation, the cognitive model explicitly postulates the existence of such negative mental schemata in "latent" form as predisposing factors to depression.

This cognitive approach to depression is interesting for several reasons. It represents a radical departure from the previously discussed intrapsychic models of depression in that it emphasizes the role that disturbances in *thinking* play in determining emotional states. It also succinctly focuses upon a central clinical feature of depression, for empirical studies have demonstrated hopelessness to be an important correlate of the depth of clinical depressive states. However, at present there is no conclusive evidence that such thinking disturbances have an etiological role in the depressive illness. Nevertheless, Beck's theory has provided the rationale for a promising psychotherapeutic approach to depression.[22] The treatment focus is on altering the negative cognitive set, with the expectation that the remaining symptoms and behaviors characteristic of depression then will be eliminated. This method of psychotherapy, it is argued, can fortify the individual against future depressive attacks by erasing the latent and predisposing negative mental schemata.

Behavioral Models of Depression

Behavioral models of depression generally attempt to focus upon characteristics of the depressed individual's immediate environment.[23] The emphasis is upon proximate events that are usually of an interpersonal or situational nature and occur in populations that may be better described as unhappy rather than melancholic. General assertions with regard to the relevance of these behavioral formulations to the genesis of melancholic mood disorder are thus difficult to make. However, because of the opportunity for objective assessment the behavioral approach offers a particular opportunity for research validation.

Depression as Learned Helplessness

Seligman, a University of Pennsylvania psychologist, has attempted to test the cognitive model of depression outlined earlier, using behavioral techniques. His formulation is based largely upon animal experiments where harnessed dogs were subjected to a series of inescapable electric shocks.[24] When subsequently these dogs were untied from their harnesses, they failed to escape the aversive stimulus, even though in their unharnessed state they were free to do so.* Only when forcibly dragged by the experimenters across a low barrier, did the dogs learn that their escape responses would bring an end to the shock experience.

Seligman extrapolates his model to human clinical depression. He suggests that a lack of assertiveness, passivity, and resignation to grim fate are learned, and are evidence of past experience where the individual has been unable to discover a behavior that successfully terminated unpleasant events. Seligman thus traces the Beckian view of helplessness to its genesis in the personal biography of the depressed patient. He postulates a trait of learned helplessness—which emerges from past states of real helplessness. The individual subsequently continues to believe that he is powerless to resolve an undesirable situation.

It is suggested that the learned-helplessness model provides an explanatory hypothesis for the predisposition to depression in passive-dependent personality. When faced with situations beyond their control, such individuals retreat to even greater passivity. Unlike the intrapsychic aggression-turned-inward model, however, the lack of assertion seen in some depressives is explained in this model as an accentuation of premorbid personality characteristics rather than retroflexion of undischarged libidinal energy.

Other behaviorally oriented clinicians have postulated that the depressive's meekness and passivity are manipulative maneuvers for obtaining interpersonal and social reward.[26] It is claimed that such reinforcement in turn shapes the depressive character structure or, in the face of a previously established depression, carries it into chronicity.

Depression as Noncontingent Reinforcement

Living creatures actively seek interaction with their environment. Should the seeking behavior (for example, the cuddling of the child by its mother) receive reinforcement (such as a quieting and signs of contentment), then the behavior is repeated. Hence the growth of any activity as preferred behavior is contingent upon that response from the environment which the behavior elicits.

In the absence of environmental reward or when the response is inappropriate or irrelevant (noncontingent) to the original behavior, the latter slowly disappears. Lewinsohn, at the Department of Psychology at the University of Oregon, has postulated that such mechanisms may underlie depressive behavior.[27]

In this model several sets of variables conspire in producing depression. Certain environments may repeatedly fail in providing rewarding opportunities, placing the individual in a chronic state of boredom, displeasure, and ultimately despair. This particular claim of the hypothesis is perhaps more relevant to social misery and unhappiness. It is further hypothesized that environmental rewards may be available, but they do not exist in a form considered relevant or important by the potentially depressed person. The bored postmenopausal housewife, whose active maternal role is lost with the departure of her children but who continues to be cared for and loved by

---

* There is considerable controversy over whether this is truly learned behavior or a deficit of central adrenergic mechanisms from the repeated shocks the dog has experienced. Experiments by Weiss et al.[25] suggest the latter. Also, the behavior disappears within approximately 72 hours of the initial experimental situation.

her husband, may become depressed if her husband's behavior has little importance for her. It is postulated also that rewards that the individual may consider undeserved lead to a further diminished self-worth. One important suggestion contained in the hypothesis is that "deficient social skills" impair or reduce the appropriate response to potentially pleasurable events, and thus decrease self-esteem.

What is not clear in the formulation is whether the diminished capacity to respond to pleasurable events is considered learned or a symptomatic manifestation of the depressive condition itself. Nevertheless, potentially the model links the individual's response to what may be the core mood experience in depression, i.e., anhedonia. It thereby provides a challenging conceptual bridge with the neurochemical substrates of pleasure and reinforcement (cf. Chapter 9).

Biological Models of Affective Disorder

Detailed attempts at biological explanation of depressive phenomena are comparatively recent. Previously, our knowledge of the nervous system was inadequate for the task. The late 1930s saw the beginnings of a functional neuroanatomy of the "emotional" brain. The limbic system is a complex network of neural interconnections that subserve the functions of drive behavior, as well as pleasure and emotion, through the motor and autonomic nervous systems. The system is intimately linked, on the one hand, to man's phylogenetically more recently acquired neocortex—which subserves symbolic and higher intellectual functions—as well as to the early, more primitive diencephalic brain centers involved in the production of hormones and the generation of biorhythmic events such as sleep and wakefulness. Upon this structural understanding have been built a complex physiology and chemistry, a brief review of which is found in Chapter 6. Current models of the biological etiology of mood dysfunction (reviewed extensively in Chapter 7) draw heavily upon these basic advances in neurobiology.

Depression as Biogenic Amine Imbalance

The hypothesis that a disturbance in biogenic amine metabolism is the predisposing factor in mood disorders grew largely from pharmacological inference in the early 1960s. Reserpine, an antihypertensive agent, was found to produce depressive depletion symptoms in about 15% of the individuals who took it. Its mode of action was the depletion of available biogenic amines. Conversely, the pharmacology of drugs found empirically to improve depression revealed that their common action by various mechanisms was to increase the presence of amines in the synaptic cleft. Several workers advanced hypotheses based on these observations—Prange (University of North Carolina)[28] in 1964; Bunney and Davis[29] and Schildkraut,[30] all from the National Institute of Mental Health, in 1965. These American workers focused almost entirely on the catecholamine system, while in Europe serotonergic mechanisms were most in favor (Coppen, 1968[31]; Lapin and Oxenkrug, 1969[32]). The pharmacological evidence was in support of both hypotheses, however, as the metabolism of both biogenic amine groups was influenced by the new antidepressants.

The biogenic amine model provides important links between clinical phenomenology and the empirical pharmacological agents that have emerged as important treatments in the affective disorders. Of all the models reviewed, it is the only one that attempts a plausible explanation for the association between manic and depressive states, suggesting that too little amine is available in depression and too much in mania. More recently, Janowsky and colleagues have extended these concepts to include cholinergic mechanisms, postulating a noradrenergic–cholinergic imbalance in affective illness.[33] While collectively these hypotheses may be recognized as a gross oversimplification, they have been an important stimulus to the development of active research programs in tbe biology of the mood disorders during the last two decades.

Endocrine Dysfunction and Mood

The clinical endocrinopathies are some of the finest experiments offered by nature in human psychopathology. Most have some association with mental abnormality. Indeed in many of the syndromes, mental derangement was among the earliest components to be described. Adrenocortical function and thyroid function have become of particular interest, because the associated endocrinopathies frequently provoke profound disturbances in mood. The possibility that hypersecretion of steroids could play a role in depression was suggested by the high incidence of depression in Cushing's syndrome and the demonstration in the late 1950s that a considerable number of depressed persons had elevated cortisol levels during their active illness. Recent studies particularly by Carroll and associates at the University of Michigan have suggested that the steroidal overproduction that shows early resistance to dexamethasone suppression may be useful as a biological marker.[34] In general, the abnormal dynamics of the steroid mechanisms suggest a limbic–diencephalic component to the depressive disturbance. Elaboration of such ideas has given rise to the hypothesis that the elevated steroids are inducing biogenic amine abnormalities or are themselves indirect evidence of biogenic amine disturbance.[32]

Similarly, disturbances in thyroid function have been suggested as modulating depressive illness, a thesis given weight by the finding that small doses of thyroid hormone added to a tricyclic antidepressant regimen increase the speed of recovery from depression, particularly in women.[35]

The Neurophysiology of Affective Disorder

Verbally retarded depressives were observed during an early study of sodium metabolism to restrict their use of optional table salt. Recognizing that sodium was a vital element in the maintenance of neuronal excitability, a major effort was mounted in the 1960s, principally in England, to define the electrolyte state in depression and mania. Coppen and co-workers of the British Medical Research Council published data that suggested increases in "residual sodium" in affective disorders.[36,37] By residual sodium is referenced a tissue compartment—including intracellular and hence intraneuronal space—which may be defined by radioisotope techniques. During an episode of clinical depression and recovery from it, redistribution of sodium appears to occur with movement in and out of the neuron, respectively. The electrolyte shift is reported to be even more pronounced in mania. The harmonious and patterned activity of the nerve cell depends on the electrical gradient across its membrane which is determined by sodium/potassium in distribution. Therefore, any excess leakage of sodium into the neuronal cell (it usually is largely outside) is expected to result in an unstable state of neurophysiological hyperexcitability.[38]

Thus, according to the electrophysiological model, both depressive and manic states are characterized by neurophysiological deviations in the same direction, with mania having the more profound disturbances in electrolyte metabolism. This view is in contradiction to the common-sense view that mania and depression are opposite conditions. Yet, it accounts for the occurrence of mixed affective states where depressive and manic features co-exist simultaneously. Furthermore, the abnormalities in the electrolyte disturbance may explain the therapeutic effects of lithium salts in both manic and bipolar depressive states.

It would be misrepresenting the integrity of the scientists actively engaged in biological research into affective disorders to suggest that any of the models outlined here purport to be a complete explanation of mood dysfunction. They do not; rather they are heuristic models and have served well in that regard, stimulating an increasing effort and enthusiasm seemingly bound only by available technologies.

In Conclusion

Some of the thinking and research that lie behind the various models we have outlined here are reviewed in later chapters. Both our conceptual integration of the information available and the extent of our knowledge have grown over the past 25 years.[39,40] Nevertheless, we still seem to categorize and broadly divide our theories between psychosocial and biological paradigms. We tend to continue to tread those divergent paths previously explored by Kraepelin and Freud. Perhaps, rather than a clever invention of Descartes', the division of mind and body has always been a natural preference of man. Descartes did indeed employ the division in a useful analytical strategy, however. We shall employ here the same device, taking up the psychological and biological approaches to mood disorder as separate entities before moving to an integration. But first we must explore the issue of classification, and the profound implications that such activity has when seeking a better understanding of our subject.

References

1. Burton, R. *The anatomy of melancholy.* New York: Vintage Books, 1977.
2. Klibansky, R., Panofsky, E., and Saxl, F. *Saturn and melancholy.* Nendeln/Liechtenstein, Kraus Reprint, 1979.
3. Lattimore, R. *Introduction to the translation of the Iliad of Homer.* Chicago: University of Chicago Press, 1951.
4. Jowett, B. *Dialogues of Plato.* London: Oxford University Press, 1892.
5. Adams, F. (Ed.). *The genuine works of Hippocrates.* Baltimore: Williams and Wilkins, 1979.
6. Summers, M. *The malleus maleficarum.* London: The Folio Society, 1968.

7. Hoffman, K. *The Print Collectors Newsletter,* 1978, *9,* 33–35.

8. Haldane, E. S., and Ross, G. R. T. (Translators). *The philosophical works of Descartes,* vol. I. Cambridge: Cambridge University Press, 1931.

9. Pinel, P. (1806). *A treatise on insanity.* New York: Hafner Publishing, 1962.

10. Esquirol, J. E. D. *Mental maladies: A treatise on insanity.* New York: Hafner Publishing, 1965.

11. Kraepelin, E. *Manic-depressive insanity and paranoia.* Edinburgh: E. and S. Livingstone, 1921.

12. Winters, E. (Ed.). *The collected works of Adolf Meyer.* Baltimore: Johns Hopkins Press, 1951.

13. Lief, A. (Ed.). *The commonsense psychiatry of Dr. Adolf Meyer.* New York: McGraw-Hill, 1948.

14. Gaylin, W. (Ed.). *The meaning of despair.* New York: Science House, 1970.

15. Abraham, K. (1911). Notes on the psychoanalytic investigation and treatment of manic–depressive insanity and allied conditions. In D. Bryan and A. Strachey (Eds.), *Selected papers on psychoanalysis.* New York: Basic Books, 1960, pp. 137–156.

16. Mallerstein, J. Depression as a pivotal affect. *American Journal of Psychotherapy,* 1968, *22,* 202–217.

17. Freud, S. (1917). Mourning and melancholia. In *Collected papers,* vol. 4. London: Hogarth Press, 1950, pp. 152–172.

18. Bowlby, J. Grief and mourning in infancy and early childhood. *Psychoanalytic Study of the Child,* 1960, *15,* 9–52.

19. Bibring, E. The mechanism of depression. In P. Greenacre, (Ed.), *Affective disorders.* New York: International Universities Press, 1965, pp. 13–48.

20. Bart, P. Depression: A sociological therapy. In P. Roman, and H. Trice, (Eds.), *Explorations in psychiatric sociology.* Philadelphia: F. A. Davis, 1974.

21. Beck, A. *Depression: Clinical, experimental and theoretical aspects.* New York: Harper and Row, 1967.

22. Beck, A. T., Rush, A. J., Shaw, D. F., Emory, G. *Cognitive therapy of depression.* New York: Guilford Press, 1979.

23. Clarkin, J. F. and Glazer, H. I. (Eds.). *Depression: Behavioral and directive intervention strategies.* New York: Garland Publishing, 1981.

24. Seligman, M. E. P. *Helplessness: On depression, development, and death.* San Francisco: W. H. Freeman, 1975.

25. Weiss, J. M., Glazer, H. I., Pohorecky, L. A., Bailey, W. H., and Schneider, L. H. Coping behavior and stress-induced behavioral depression: Studies of the role of brain catecholamines. In R. A. Depue (Ed.), *The psychobiology of the depressive disorders.* New York: Academic Press, 1979, pp. 125–160.

26. Ferster, C. Classification of behavior pathology. In L. Krasner, and L. Ullman, (Eds.), *Research in behavior modification.* New York: Holt, Rinehart and Winston, 1965, pp. 6–26.

27. Lewinsohn, P. A behavioral approach to depression. In R. J. Friedman and M. Katz (Ed.), *The psychology of depression: Contemporary theory and research.* Washington: U.S. Government Printing Office.

28. Prange, A. The pharmacology and biochemistry of depression. *Diseases of the Nervous System,* 1964, *25,* 217–221.

29. Bunney, W. E., Jr., and Davis, M. Norepinephrine in depressive reactions. *Archives of General Psychiatry,* 1965, *13,* 483–494.

30. Schildkraut, J. Catecholamine hypothesis of affective disorders. *American Journal of Psychiatry,* 1965, *122,* 509–522.

31. Coppen, A. Depressed states and indolealkylamines. *Advances in pharmacology,* vol. 6. New York: Academic Press, 1968, pp. 283–291.

32. Lapin, I., and Oxenkrug, G. Intensification of the central serotonergic process as a possible determinant of thymoleptic effect. *Lancet,* 1969, *1,* 132–136.

33. Janowsky, D., El-Yousef, K., Davis, M., and Sekerke, H. J. A cholinergic–adrenergic hypothesis of mania and depression. *Lancet,* 2, 1972, 632–635.

34. Carroll, B. J., Feinberg, M., Greden, J., Tarika, J., Albala, A. A., Haskett, R. F., James, M. N., Kronfol, Z., Lohr, N., Steiner, M., de Vigne, J. P., and Young, E. A specific laboratory test for the diagnosis of melancholia. *Archives of General Psychiatry,* 1981, *38,* 15–27.

35. Prange, A. J., Wilson, I. C., Rabon, A. M., and Lipton, M. A. Enhancement of imipramine antidepressant activity by thyroid hormone. *American Journal of Psychiatry,* 1969, *126,* 457–469.

36. Coppen, A., and Shaw, D. Mineral metabolism in melancholia. *British Medical Journal*, 1963, *2*, 1439–1444.
37. Coppen, A., Shaw, D., Mallerson, A., and Costain, R. Mineral metabolism in mania. *British Medical Journal*, 1966, *1*, 71–75.
38. Whybrow, P., and Mendels, J. Toward a biology of depression: Some suggestions from neurophysiology. *American Journal of Psychiatry*, 1969, *125*, 45–54.
39. Akiskal, H. S., and McKinney, W. T., Jr. Depressive disorders: Toward a unified hypothesis. *Science*, 1973, *182*, 20–28.
40. Whybrow, P., and Parlatore, A. Melancholia, a model in madness: A discussion of recent psychobiologic research into depressive illness. *Psychiatry in Medicine*, 1973, *4*, 351–378.

3

Clinical and Familial Subtypes of Mood Disorders: Observation, Opinion, and Purpose

The classification, or avoidance of classification, which I would propose is: first, to determine whether the depression is severe or mild—melancholia or neurasthenia, descriptively speaking; and, of course, there are many grades. Next . . . on what discoverable influences does it depend: age; constitutional morbid trends; lifelong environmental factors, recent more acutely disturbing ones? What, in sum, is the presumptive balance of environmental forces responsible for the illness as against inherited ones. . . . Has the morbid condition . . . established an autonomy? Have the patient's character and surroundings worked to his advantage or disadvantage? These are points to weigh rather than points on which to classify . . . because we have no sure means of distinguishing exactly the numerous cases in each case. . . . No doubt increasing knowledge will bring an improved . . . classification system based on etiology . . . whether it comes by way of genetics, psychology, or somatic psychology, it will be welcome so long as it is useful and valid.

Sir Aubrey Lewis[1] (1938)

Why Classify?

Classification has the general purpose of promoting understanding. In science, the ordering of information seeks to make past observation more intelligible and future events more predictable. It also acts as an implicit guide to those who are conducting research into various phenomena. Principally, the strategies employed in classification are the clustering of like phenomena into various groups or distributing them upon a continuum or dimension of change. Observation of how relationships among the phenomena change over time adds a sophistication to either strategy. Charles Darwin devoted his life to one grand classification—to the development of a systematic arrangement by which all living and extinct organisms are united by complex, radiating, and circuitous lines of affinities into a few grand classes.[2] He rested his theory of the origin of the species upon the meticulous clustering of the wide variety of phenomena that he observed during his voyage with the "Beagle."

In medicine, making a diagnosis is a form of classification. It is the clustering or distribution of disorders based upon their phenomenology and what we know of their natural history. Whereas in science in general the ordering seeks to improve conceptual understanding and to guide further investigation, in medicine there is an intense need to go beyond this fundamental principle and to seek a classification that will assist in predicting the outcome of intervention. The reason is obvious, for it is the natural course of a disorder and the response to specific treatments that are of central concern to both patient and physician. Diagnosis is thus a very practical form of classification.

Therapeutic intervention and advice to a patient on how to meet the challenge of a particular illness are ideally based upon a knowledge of pathogenesis—of the fundamental etiology of the disorder. Such knowledge increases our predictive ability manyfold. It thus makes our recommendations more valid. Unfortunately, however, such a sophisticated understanding is attained in medicine only rarely. This has been especially so in disorders of behavior. Indeed, some have argued that the contributions to the genesis of mental dysfunction are so varied and the disorders themselves so diverse, perhaps each unique to the individual, that the exercise is a useless one. Fortunately, most behavioral scientists do not subscribe to such nihilism.

Phenomenology and Pathogenesis: Confusion and Debate

In the absence of a precise knowledge of pathogenesis, opinion varies regarding the meaning of the phenomena observed and tends to distort the way in which those phenomena are classified. Although one may agree with the aphorism of John Milton that "opinion in good men is but knowledge in the making,"[3] it has proved to be wise in science to keep observation and opinion distinct. As the reader will have concluded from the historical background to our discussion of mood disorders in the last chapter, such a distinction in disorders of behavior has been hard to achieve. A preconception regarding the origin of mental derangement justified inhumane treatment of insane persons until very recent times. Similar perceptions have confounded efforts to achieve an objective classification of mental disorders, including mood disorder.*

An interesting, comparatively recent, example of confusion between the purpose of diagnosis and opinion regarding etiology has been the predominantly English debate whether melancholic ("endogenous") depressions are fundamentally different from those depressions that appear as psychologically understandable responses to life's events. Confusion about syndrome description and therapeutic outcome, due to differing opinions regarding pathogenesis, appears to be at the core of the argument. One school of thought maintains that both psychological and biological factors are relevant to the etiology at all forms of depression. Sir Aubrey Lewis of the Maudsley Hospital,

---

* The Task Force on Nomenclature and Statistics of the American Psychiatric Association, the group which oversaw the Third *Diagonostic and Statistical Manual* of the Association (*DSM-III*),[4] was acutely aware of these difficulties. A very special effort to base the classification on phenomenology was made as a result.

London, advised weighing the relative contributions of psychological and physical factors in the psychogenesis of the individual case and avoiding categorical classification (see our chapter quotation).[1] He believed that the phenomena observed could best be ranged upon a continuum—a symptom dimension. This unitarian (perhaps Meyerian) position was opposed by another group of psychiatrists in Newcastle, England. This group maintained that on the basis of both observation and treatment outcome, there existed two distinct types of depression. To one of these they imputed a genetic or biochemical basis (psychotic or endogenous depression), while the other (usually termed neurotic or reactive), differing only quantitatively from normal disturbances of mood, was environmentally induced.[5]

Whether the phenomena, when objectively described, support a dichotomy or a continuum with regard to therapeutic response is not the root of the confusion. Rather, it is the imputing of pathogenesis as either biological or environmental and the inference that either biological or psychological treatment would therefore be beneficial that led to the heated debate and a potential distortion of objective therapeutic management.

Lewis, in fact, emphasizing that the purpose of any classification is to be useful and valid, had tried to avoid this conceptual pitfall by clustering the phenomena into what Kendell termed Type A and Type B depressions.[6] These groups could be considered either as two distinct forms or as representing the extreme poles of a normal distribution. "Type A" referred to acute severe depression with a diurnal variation of mood, guilt, and significant dysfunction in vegetative and psychomotor activity. "Type B" referred to a residual group of milder and more chronic depressions that lacked such features, but were characterized by prominent neurotic character pathology.

The English debate is the most celebrated of many similar discussions. Various investigators, beginning with Moebius in the 19th century, have proposed comparable, though not necessarily synonymous, dichotomous schemes of classification, e.g., endogenous–exogenous, endogenomorphic–nonendogenomorphic, autonomous–reactive, psychotic–neurotic, vital–personal, major–minor, and physiological–psychological. The heated exchanges perhaps reflect a special example of the mind–body dichotomy that has characterized the Western approach to the understanding of all human illness. Alternatively, could they have their roots in the ancient rivalry between the Platonic and Hippocratic conceptions of disease? The Platonic school postulated the real existence of pure categories or types of illness; the purpose of all philosophical or scientific inquiry was to discover such ideal forms. The Hippocratic approach, on the other hand, rooted in an empirical tradition, emphasized observation and description on actual cases of disorder to be understood as variations from the normal state of health. Hippocratic physicians thus focused their efforts on what we now term a dimensional classification of human diseases.

It is important to recognize that the two approaches are not incompatible or even in competition: the Platonic or categorical approach is a convenient scientific strategy when first studying a disorder in which knowledge of pathogenesis is uncertain and its boundaries thus need to be sharply defined to avoid confusion with other conditions.[7,8] As more is learned about the illness, clinicians become more confident in reliable identification of it and are usually willing then to concede that many transitional forms between the "normal" and the diseased states do exist. Then with the observed phenomena accurately described and agreed upon, an objective classification which incorporates the value of drawing upon both the categorical and dimensional approach begins to take form. With the new *DSM-III* classification we are now arriving at such a point in the evolution of the nosology of mood disorder. A clear distinction between that which is observed and that which is imputed regarding etiology has been made the cornerstone of the classification. While concepts of cause are essential in guiding new directions in research, they can also lead to circuitous argument, as in the great English debate, when opinion regarding etiology is confused with explanation.

Symptomatology and the Classification of Mood Disorders

Moebius' categorical distinction between exogenous and endogenous mental disorders during the 19th century was an important advance for practical psychiatric nosology. It separated mental disorders that could be accounted for by discoverable *external* causes such as infectious agents from those in which none could be found. For the latter group, *internal*, i.e., constitutional or hereditary, factors were presumed. These concepts, applied to affective disorders by Kraepelin and his followers, led to the clustering of all major disorders of mood into one class, that of manic–depressive insanity. Kraepelin considered environmental factors of little consequence in the core illness.

> The certain conclusion, which can be drawn from these and similar extremely frequent (environmental) experiences, leads us to this, that we must regard all alleged injuries as possible sparks for the discharge of individual attacks, but that the real cause of the malady must be sought in *permanent internal changes*, which at least very often, perhaps always, are innate. Unfortunately the powerlessness of our efforts to cure must only too often convince us that the attacks of manic–depressive insanity may be to an astonishing degree independent of external influences.[9]

Subsequently, many of the proposed subclassifications of Kraepelin's original clustering have returned to the fundamental division proposed by Moebius. There is undoubtedly a clinical appeal to an approach that makes a categorical distinction between psychological and hereditary factors in pathogenesis. However, the dichotomy is a dangerously simple one. It is increasingly clear that both environmental and hereditary factors contribute to the pathogenesis of all disease states. Although some genes may "penetrate" to cause disease in almost all environments, most genes require a specific environment for their expression. Hence "exogenous" factors are, in all probability, a necessary contribution to the unfolding of "endogenous" processes in the genesis of most diseases. Reciprocally, it is difficult to imagine how an "exogenous" factor could cause a psychiatric disorder without reference to some inherent "endogenous" characteristic of the human organism.

Furthermore, one of the cardinal predictions of the exogenous–endogenous dichotomy in affective disorders—i.e., absence of psychosocial precipitants on endogenous depressions—is not supported by current evidence. For instance, Thomson and Hendric could not demonstate an inverse relationship between stress scores measured on the Holmes–Rahe scale and family history for affective illness taken as an indirect measure of hereditary predisposition.[10] Also, Leff and associates at the National Institute of Mental Health, Bethesda, reported that the absence of a history of psychosocial precipitants in endogenously depressed persons was secondary to the severity of the illness. Such patients were too disturbed to appraise the psychosocial context of their illness. Upon recovery, however, these individuals were as able as "nonendogenous" patients to identify psychosocial factors that preceded the onset of their depression.[11] More recently Hirschfeld, also at the National Institute of Mental Health, failed to validate the existence of a separate category of "situational depression" distinct from other forms of depression, defined by the presence of significant psychosocial precipitants.[12]

Hence, although it would appear that the categorical clustering of depression by symptom association into "endogenous" and "exogenous" forms may be valid and useful in predicting clinical course, such a distinction cannot reasonably bear the weight of cause. If we relieve the dichotomy of this unnatural burden, however, the emphasis is constructively shifted from a clustering based on putative etiology to a clinical classification based on phenomenology. Then, from the standpoint of a practicing clinician, the classification potentially becomes both useful and valid.

The Need to Be Practical

From the practical standpoint, careful analysis of quantitative change along a single symptom dimension can lead to the formation of specific categories or classes of phenomena. Both dimensions and categories are methods of empirical organization. We *choose* to create the discontinuity between diagnostic categories by defining the degree of quantitative change that will be accepted before a qualitative difference is presumed to have occurred. As there are many *types* of information to be considered—symptoms, signs, past history, family history, environmental challenge, laboratory challenge (e.g., psychological tests, dexamethasone suppression), previous response to treatment—there also will be many cross-cutting continua to be assembled and assessed. It is not possible to categorize all the phenomena, and thus any classification will be imperfect and open to legitimate criticism. If the classification does not include the clinically significant syndromes or patterns, however, and does not have some validity in defining outcome and therapeutic response, in the practical world it will be eventually discarded. Thus, one finds that those major elements of a classification that survive usually have significant and practical value.

A Primary–Secondary Distinction

Not all individuals who suffer mood disorder experience the classical syndromes described in Chapter 1. There are intermediate and transitional forms, and frequently the affective disturbance is associated with other illness. The primary–secondary distinction proposed by Robins and Guze at Washington University, St. Louis, was devised to meet this latter complexity.[13] "Secondary depression" is defined as a mood disorder *chronologically* superimposed on a preexisting nonaffective disorder; the latter includes incapacitating physical disorders, as well as the 12 nonaffective psychiatric conditions recognized by the St. Louis investigators to be valid nosological entities.* "Primary depressions," by contrast, arise in the absence of such preexisting disorders. That a nonaffective disorder chronologically precedes the depressive state does not imply that the former represents the etiological substrate for the latter, however. The principal aim was to define a "pure" population of depressed persons (primary depressives) *uncontaminated* by other illness, who could then be objectively studied.

The longitudinal validity of this primary grouping was investigated by Murphy and associates, who showed that a representative cohort of primary depressives displayed remarkable diagnostic stability over a 5-year period of prospective follow-up.[16] Patients diagnosed as primary depressive at first (index) interview did not develop other psychiatric disorders over a period of 5 years. By contrast, secondary depressions, reported in a series of papers by the St. Louis group, resembled more the underlying nonaffective disorders than primary depressions when examined over time.[17,18]

Alcoholism, schizophrenia, and sociopathy appear to be the most common primary psychiatric diagnoses to which depression is secondary in the inpatient setting. However, Akiskal and associates in a University of Tennessee study of "neurotic" depression found that, in the outpatient setting, secondary depressions are more evenly distributed among all common major psychiatric conditions. In addition to the three disorders mentioned, these include such conditions as Briquet's hysteria, anxiety, phobic and obsessive–compulsive disorders, and (ego-dystonic) homosexuality.[19]

The Memphis study also reported that the clinical picture in secondary depressions was often nonsyndromal, with prominence of the "subjective" rather than the "objective" manifestations of depression. Hypomanic responses to tricyclic challenge did not occur, and family history was similar to that found in nonaffective states.[20] As for biological differentiation of these secondary depressions from primary mood disorders, the REM latency findings of Kupfer and Thase from the University of Pittsburgh constitute a promising marker.[21] The latency of the first period of rapid eye movement (REM) from the first onset of sleep with Stage II, which is normally 70–110 minutes, is somewhat reduced in secondary depressives, and ranges from 15–60 minutes in primary depressive disorders. The dexamethasone suppression test may similarly differentiate between primary and nonsyndromal secondary depressions. These findings have been recently replicated in an outpatient sample of depressives studied by Akiskal, Lemmi and colleagues in Memphis.[22] Some of the clinical and biological differences between primary and secondary depressions are summarized in Table 3.1.

The therapeutic validity of the distinction is less clear. In the absence of controlled studies to compare the response of the two groups to tricyclic drugs and electroconvulsive therapy, it is uncertain whether secondary depressions are less likely than primary cases to respond to these treatment modalities. Also, inpatient studies have reported

---

* Schizophrenia, sociopathy, Briquet's syndrome, anxiety, phobic and obsessional disorders, alcoholism, drug dependency, organic brain syndromes, mental retardation, anorexia nervosa, (ego-dystonic) homosexuality, and transsexualism.[14,15]

Table 3.1 Reported Clinical and Biological Differences between Primary and Secondary Depressions[a]

	Primary	Nonprimary (secondary)
Definitional (preexisting non-affective disorder)	No	Yes
Symptomatological	Signs as prominent as symptoms ("objective" depression)	Symptoms more prominent than signs ("subjective" depression)
Perception of illness by patient	As a definable break in personality	Accentuation of usual complaints
Course	More episodic than chronic	More chronic than episodic
Prognosis	Usually good	Depends on underlying disorder
Sleep EEG: REM latency	Shortened	Within lower range of normal

[a]From reference 20.

alcoholism to be the most frequent "primary" condition in secondary mood disorders.[18] Here the primary–secondary distinction may break down. For instance:

1. It is quite probable that some forms of alcohol abuse represent early symptomatic manifestations of mood disorder.[23]
2. Transient depressive symptoms commonly occur in alcoholics undergoing detoxification, i.e., when most alcoholics seek medical or psychiatric attention. Keeler et al. have convincingly argued that such dysphoric manifestations do not constitute an affective disorder.[24]
3. Another possibility is that, because of assortative mating between an alcoholic and an affectively ill parent (a common pattern of marriage, according to Dunner and co-workers[25]), the individual may suffer from two separate illnesses. Alcoholism, which usually has early onset, will often precede the affective disorder in such cases, again leading to the ambiguous relegation of the mood disorder to "secondary" category.

Many of the critical points regarding the coexistence of alcoholism and depression apply to depressions occurring in the setting of serious medical illness. At one level, the incapacitation and the limitations that are imposed on the sufferer's life style produce sufficient demoralization to be confused with a depressive disorder. This is the principal reason why Robins and Guze first introduced the concept of depression secondary to incapacitating medical disorder. However, like alcohol withdrawal, a medical disorder may cause many nonspecific symptoms of depression—such as insomnia, anorexia, fatigue, psychomotor slowing, and poor concentration—that together with demoralization might be construed as evidence of a genuine mood disorder. Further, either the somatic process itself or the medications used for its management may alter the neurochemical substrates of mood and give rise to an unmistakable melancholia (or even mania). Finally, a patient may suffer from two separate but interrelated diseases. A medical illness, for example, may unmask the genetic potential for a depressive disorder. In the primary–secondary scheme, individuals in all four categories are likely to receive the designation of "secondary depression."

Such complexities may explain why there have been conflicting results in studying the primary–secondary distinction. It has not received universal acceptance among researchers. However, the effort to identify a group of individuals with relatively "uncontaminated" depression is of therapeutic, heuristic, and research importance, allowing the dimensions of major affective illness to be clarified uncomplicated by other psychiatric and physical disorders. This has been a critical step in objective classification of "pure" affective illness.

Classifying Primary Mood Disorder: Dimensional and Axial Techniques

It is not necessary to agree upon etiology or pathophysiological process to objectively describe and define a syndrome or cluster of phenomena that appear to have natural association. Ideally, definitions are arrived at by consensus among clinicians and researchers and they change as their validity to predict therapeutic outcome, natural history, genetic vulnerability, and other clinically significant factors is determined. In disorders such as those of mood, in which the magnitude or dimensions of the phenomena vary quantitatively with time, the variations themselves can be used to organize the information. From these dimensions categories can be agreed upon. For example:

- From the dimension of mood: a unipolar–bipolar category
- From the dimension of time: an acute–chronic category
- From the dimension of severity: a psychotic–nonpsychotic category
- From the dimension of the apparent response of the disorder to environmental intervention: the concept of autonomy

The *Diagnostic and Statistical Manual (DSM-III)* utilizes some of these concepts in the definition of the major psychopathological syndromes.[4] In addition, it employs a multiaxial evaluation technique. This allows each individual case to be diagnostically reviewed not only with regard to the symptoms and signs of the primary disorder (Axis I) but also to several other classes of information which bear upon the principal illness:

Axis II: Long-standing patterns of personality and/or developmental dysfunction

Axis III: Physical disorders and conditions (A partially overlapping method for handling this class of information has already been discussed in the primary–secondary dichotomy.)

Axis IV: Severity of psychosocial stressors

Axis V: Adaptive function in previous year

Using such a method, a larger volume of useful information can be organized and communicated. Some of this additional information may emerge eventually as of predictive value.

At present, phenomenology (in *DSM-III*, Axis I) remains the most important component of any psychiatric classification. By careful description and agreed-upon definition, the major elements are clustered into categories. In creating these clusters or categories, it is necessary for researchers and clinicians to develop intelligent but nonetheless arbitrary cut-off points to divide up the natural continuum. Sometimes these divisions are made easier by clusters sharing a similar pattern of therapeutic response or family illness.

The recent definition of the unipolar–bipolar dichotomy in mood disorders is an excellent example of the process involved.

Family Studies and the Unipolar/Bipolar Distinction

The unipolar–bipolar distinction, originally proposed by Leonhard,[26] directly challenges Kraepelin's concept of manic–depressive insanity. Specifically, it utilizes the differences in mood found in the disorders to develop categories that have validity in relation to the clustering of these differences in families.

The observation that affective disorders tend to cluster in families is not new. Kraepelin, among those patients he observed in Heidelberg, believed he could demonstrate a "hereditary taint" in about 80% of the individuals with manic–depressive insanity. However, it was the thrust of Kraepelin's thesis that mania, melancholia, and combinations of the two disorders in the same individual were all "manifestations of a single morbid process." He cited his observations of families in support of his viewpoint:

> The various forms . . . may also apparently replace one another in heredity. In members of the same family we, frequently enough, find side by side pronounced periodic or circular cases, occasionally isolated states of ill temper or confusion, lastly very slight, regular fluctuations of mood or permanent conspicuous coloration of disposition.[9]

As our understanding of the patterns of inheritance has advanced, family studies have become an important element in clarifying the biological permission that may exist for some individuals suffering affective disorder. In the course of such studies in the early 1960s, Leonhard noted that in contrast to those of individuals with depressed mood alone, the relatives of an individual with a history of mania, whether in isolation or in association with severe depression, were at greater risk for affective illness. He suggested a new subclassification, using the term *bipolar* for index persons with mania and *monopolar* (later termed *unipolar*) for those with depression alone. Furthermore, he concluded that mania, if used as an organizing variable in classification, supported the presence of genetically distinct categories of mood disorder—one variety associated with mania (bipolar illness) and one without such an association (unipolar disorder).

Angst,[27] Perris,[28] and Winokur et al.,[29] in a large series, supported Leonhard's findings and observed further differences—that bipolar probands tend to have both bipolar and unipolar relatives, whereas unipolar individuals have largely unipolar offspring.

From the standpoint of classification and validity the unipolar–bipolar distinction is an important one, because it hinges upon phenomena that can be relatively clearly defined. Broadly speaking, elation is not difficult to distinguish from depression. Hence it has been possible to obtain reasonable support for the fundamental principle of Leonhard's classification. From the practical standpoint, agreement on the *degree* of mood change and associated features that are necessary to define the qualitative change from euthymia to hypomania and mania was essential. Table 3.2 gives the *DSM-III* definition of a manic episode as an example. Similarly, it has been necessary to agree upon the criteria for a major depressive episode (Table 3.3). The number of such episodes *without* mania occurring that must be apparent before bipolar illness can be reasonably ruled out must also be agreed upon. The Research Diagnostic Criteria (RDC)[30] and more recently the *DSM-III*[4] have sought to develop agreement on these distinctions, enabling new information to be gathered and organized. They have become useful working tools both in clinical practice and research. Table 3.4 (see page 54) summarizes familial, clinical course, and pharmacological response characteristics that have been reported by researchers in support for the unipolar–bipolar distinction;[31–33] not listed on the table are nonreplicated biological findings such as differences in platelet monoamine oxidase activity, steroid output, and average evoked potentials in the EEG.[34] Much information has been gathered, and thus the two categories have served a useful purpose. However, the cautionary note struck earlier must again be emphasized. As clusterings of phenomena, they are not perfect but useful tools. Modification is to be expected as experience with the categories accumulates. Many important questions regarding the unipolar–bipolar dichotomy are still unsettled. For instance:

- What *degree* of elation should be used as a cutoff point between unipolar and bipolar illnesses? There appears to exist a large universe of "transitional"

Table 3.2 *DSM-III* Diagnostic Criteria for Manic Episode

A. One or more distinct periods with a predominantly elevated, expansive, or irritable mood. The elevated or irritable mood must be a prominent part of the illness and relatively persistent, although it may alternate or intermingle with depressive mood.
B. Duration of at least 1 week (or any duration if hospitalization is necessary), during which, for most of the time, at least three of the following symptoms have persisted (four if the mood is only irritable) and have been present to a significant degree:
  1. Increase in activity (either socially, at work, or sexually) or physical restlessness
  2. More talkative than usual or pressure to keep talking
  3. Flight of ideas or subjective experience that thoughts are racing
  4. Inflated self-esteem (grandiosity, which may be delusional)
  5. Decreased need for sleep
  6. Distractibility, i.e., attention too easily drawn to unimportant or irrelevant external stimuli
  7. Excessive involvement in activities that have a high potential for painful consequences which is not recognized, e.g., buying sprees, sexual indiscretions, foolish business investments, reckless driving

Table 3.3 *DSM-III* Diagnostic Criteria for Major Depressive Episode

A. Dysphoric mood or loss of interest or pleasure in all or almost all usual activities and pastimes. The dysphoric mood is characterized by symptoms such as the following: depressed, sad, blue, hopeless, low, down in the dumps, irritable. The mood disturbance must be prominent and relatively persistent, but not necessarily the most dominant symptom, and does not include momentary shifts from one dysphoric mood to another dysphoric mood, e.g., anxiety to depression to anger, such as are seen in state of acute psychotic turmoil. (For children under 6, dysphoric mood may have to be inferred from a persistently sad facial expression.)

B. At least four of the following symptoms have each been present nearly every day for a period of at least 2 weeks (in children under six, at least three of the first four):

   1. Poor appetite or significant weight loss (when not dieting) or increased appetite or significant weight gain (in children under six, consider failure to make expected weight gains)
   2. Insomnia or hypersomnia
   3. Psychomotor agitation or retardation (but not merely subjective feelings of restlessness or being slowed down) (in children under 6, hypoactivity)
   4. Loss of interest or pleasure in usual activities, or decrease in sexual drive not limited to a period when delusional or hallucinating (in children under 6, signs of apathy)
   5. Loss of energy, fatigue
   6. Feelings of worthlessness, self-reproach, or excessive or inappropriate guilt (either may be delusional)
   7. Complaints or evidence of diminished ability to think or to concentrate, such as slowed thinking, or indecisiveness not associated with marked loosening of associations or incoherence
   8. Recurrent thoughts of death, suicidal ideation, wishes to be dead, or suicide attempt

forms of mood disturbance between the classical definitions of unipolar and bipolar disorders.[35,36] The acknowledgement of such transitional clinical forms* emphasizes the dimensional nature of mood and its disorders.

- How many *episodes* of depression without elation should occur before an individual is considered to be a unipolar proband? The risk of hypomanic or manic episodes is not negligible, ranging from 4%–33% in different series of depressed persons examined prospectively. It is especially high in recurrent depressions, and may occur as late as 10 or more years after onset of the first depressive episode.

- What are the *respective clinical boundaries* of unipolar and bipolar illnesses? It has been known since Kraepelin's time—and recently confirmed by Gershon and associates' genetic studies in Jerusalem,[37] Akiskal and colleagues in Memphis,[23] and Depue et al. in Buffalo[38]—that attenuated cyclothymic forms of bipolar illness exist. Whether subaffective dysthymic forms of unipolar illness exist is less clearly established, but the work of these investigators suggests that at least some forms of so-called "minor" or "characterological" depressions also belong to primary affective disorder. Thus, a clinical focus on major affective illness alone may provide homogeneous affective populations for research comparison, but it also *restricts testing the hypothetical unipolar–bipolar distinction by presuming its validity.* Akiskal, for example, has noted a

---

* Variously referred to as Bipolar II (retarded depressions with brief and adaptive hypomanic periods) and Bipolar III (recurrent depressions with bipolar family history and/or brief pharmacologically induced hypomanic periods).

Table 3.4 Definitional, Familial, Clinical, and Pharmacological Differences between Unipolar and Bipolar Depressions[a]

	Unipolar	Bipolar
Definitional		
History of hypomania or mania	No	Yes
Familial		
Family history	Unipolar and alcoholism	Unipolar Bipolar Alcoholism?
Morbidity risk of affective illness in biological relatives	Lower	Higher
Two to three consecutive generation pedigrees	Uncommon	Common
Familial risk for diabetes, hypertension, and coronary artery disease	No	Modestly increased?
Natural history		
Phenomenology	Anxiety, agitation, somatic complaints	Hypersomnia, psychomotor retardation
"Typical" age of first episode	30s to 60s	Teens to 30s
"Typical" premorbid personality	Any type	Cyclothymic, or obsessoid
Postpartum episodes	Less common	More common
Course	Few episodes per lifetime	Multiple episodes per lifetime
Length of episodes	6–9 months	3–6 months
Marital failure	Less common	More common
Pharmacological		
Tricyclics	Responsive	Less responsive, may switch to hypomania
L-Dopa	No hypomanic excursion	Regular induction of hypomania
Lithium	No acute antidepressant effect, useful in the prophylaxis of selected cases	Modest antidepressant effect, prominent antimanic and prophylactic effect

[a]From references 31–33.

high frequency of hypomanic responses to tricyclic challenge in a subgroup of dysthymic individuals.[39] Such responses were uncommon in the reference population of classical unipolar persons.

These observations again emphasize the continuum of clinical phenomena between unipolar and bipolar categories. Hence, although the categories have served well in advancing our understanding, if we entirely confine our clinical and research attention to them, further advances may be blocked. Clinical classification—diagnosis—is essential to practical therapeutic management. However, the distinctions we make must be accepted in that context and also be open to change when new evidence emerges. When categories emerge as catechism, inevitably reasonable doubt and inquiry decline.

Other Specific Disorders: Variants of the Major Affective Disorders

The existence of disabling illness that lies on the fringe of the major affective syndromes has been increasingly recognized in recent years. While originally it was presumed that such disorders were "neurotic" or minor and largely environmentally induced (cf. the English debate), that is now generally recognized as simplification. Whether one designates these dysfunctions as "other specific disorders," which is the term used in *DSM-III*, or as Type B depressions, as did Aubrey Lewis, it is clear that we need to learn more about them.

Most of the depression experienced in these categories tends to be seen in the office of the primary physician or in the hospital outpatient clinics and mental health clinics. Rarely in the first instance are these individuals hospitalized. Akiskal, Bitar, and colleagues undertook a comprehensive investigation of a cohort of 100 such persons with "ambulatory depression" in Memphis.[19] In the initial phase of that study, an attempt was made to clarify the *operational* meaning of the term "neurotic" as applied to depression. As may be judged from the following responses, it was found that clinicians used the concept broadly with much personal variation.

- *Mild Illness*—where a depressive state falls short of the criteria for the full syndrome, does not incapacitate the individual, and ambulatory care is usually possible
- *Nonpsychotic*—affective manifestations that do not acquire delusional or hallucinatory quality
- *Nonendogenous*—only mild vegetative signs without the characteristic anhedonia of severe illness; the mood disturbance is influenced by environmental events
- *Coexistence of neurotic symptoms* such as phobic, obsessional, and hysterical manifestations
- *Reactive*—the depression can be understood in the light of antecedent psychological circumstances, most commonly losses
- *Characterological*—a lifelong tendency to overreact to "normative" stress by developing dysphoric symptoms. A liberal mixture of "unstable" character traits such as passive-dependence, immaturity, manipulativeness, impulsivity, low frustration tolerance, and proneness to suicidal gestures is common

Such widely divergent differences in interpretation of a "diagnostic" term make the sharing and comparing of clinical information difficult. Objective studies of outcome are frequently helpful in such situations. Hence in the second phase of the study the clinical course of the cohort was followed prospectively over a three- to four-year period. Ninety-seven percent of the individuals were assessed by direct personal interview. The results, summarized in Table 3.5 (see page 56), were somewhat unexpected. Most importantly, 40% of these originally "minor" depressions developed phenomena characteristic of the "major" disorders. Melancholic, "hypomanic," "manic," and "psychotic" episodes all occurred. Serious character pathology was found equally distributed in both "minor" and "major" depressions, but completed suicide (three cases) was limited to the former. An unfavorable social outcome was also more common in the former group. These data are consistent with the findings of Paykel and

Table 3.5 Three- to Four-Year Outcome in
Neurotic Depressions[a]

	Number ($N = 100$)[b]
Manic episode	4
Hypomanic episode	14
Psychotic depression	21
Melancholic depression	36
Episodic course	42
"Characterological" features	24
Social invalidism	35
Suicide	3

[a]From reference 19.
[b]Sum total more than 100 since more than one outcome was
possible per patient.

co-workers in New Haven,[40] Pilowsky et al. in Sheffield,[41] and Klerman and associates in the National Institute of Mental Health collaborative study.[42] The conclusion is, of course, that "neurotic" depression is not a useful nosological entity. Other important elements, some of which we have noted earlier, also emerged from the findings[19]:

- The presence of *psychosocial precipitants is of neutral value* in the classification of affective disorders.
- *"Neurotic" depressions are heterogeneous,* a substantial proportion forming the precursors of major affective disorders.
- *Severity of illness is hard to define* and thus does not distinguish well. Some dysthymic depressions had such serious complications as suicide. Furthermore, major depressions tended to be more episodic than chronic and frequently had a better social outcome than the minor so-called neurotic depressions.
- Serious *character pathology also does not distinguish.* Therefore, as endorsed by *DSM-III*, it is preferable to code character disorder on a separate axis (Axis II) orthogonal to the phenomenological affective axis (Axis I).

Many "other specific affective disorders" run a chronic course. Indeed, this appears to be a distinguishing characteristic from the major depressive syndromes, which tend to be episodic and interspersed with relative euthymia.

Toward a Practical Nosology: DSM-III

Whatever classification of mood disorders is adopted, the divisions will remain in debate until we have a more precise knowledge of pathophysiology. As we have argued, this is not a reason for dismissing the discipline of classification, however, for there are many research and therapeutic benefits to such an exercise.

In fact, it is the improvement and growing diversity of the treatment of mood disorder that makes our modern effort at subclassification meaningful. The response to therapy indeed becomes a variable against which the classification can be validated.

This validation was kept very much in mind during the development of *DSM-III*. [4] Both existing research on validation and a structure that would lend itself to future research were taken into account by the task force. Of particular importance was the development of clear, pharmacologically based diagnostic criteria that could be applied by clinicians and researchers of varying theoretical persuasions. Specifically with regard to the classification of affective disorders, many of the issues that we have discussed in this chapter were considered and incorporated into Axis I of the new nomenclature. While some disagreement remains regarding details, a substantial consensus was reached which can now form the basis for clinical management and further research.

The *DSM-III* classification (Axis I) is outlined in Fig. 3.1. In using the classification we have found helpful the following series of steps.

When first confronted with an individual who complains of mood disturbance and has prominent affective symptoms, *the chronology of the disturbance in relation to other illness* should be considered. Does this disturbance emerge de novo, without previous psychiatric or physical disorder? Current evidence indicates that transient or even low-grade intermittently chronic depressive symptoms do occur in the course of many nonaffective psychiatric disorders.[39] In most situations such dysphoria is an integral part of, or a reaction to, the underlying disorder. Should the syndrome be *secondary* to other illness, then, rather than confounding the phenomenology of mood

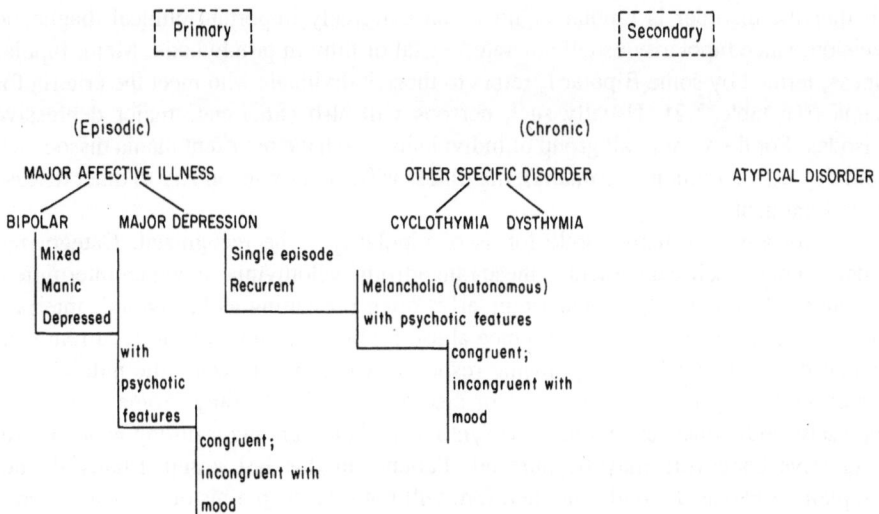

Figure 3.1 Some elements of Axis I of the *DSM-III* classification. *DSM-III* does not utilize the primary/secondary distinction that we have found useful from a research standpoint.

disorder, it is recommended that the disturbance be noted on a separate axis. This is an important feature of the new diagnostic manual. Axis III is reserved for physical disorders and conditions, and Axis II for personality disorder of long standing or developmental disorder. Information in both categories may be of great relevance to the dominant mood disorder. The medical condition may have etiological significance (e.g., hypothyroidism) or may suggest or limit the somatic treatment choices (as is the case with hypertension or coronary artery disease).

Similarly, any significant personality disorder clearly antedating the affective episode should be coded on Axis II. Previously, characterological features considered chronic and unstable—e.g., passive-dependence, histrionic features, sociopathy— were thought to suggest the diagnosis of a "neurotic" depression, while "stable" personality traits such as obsessionalism or narcissism were believed to be more commonly associated with "endogenous" depressions. According to current evidence, clearly demonstrated in the Memphis studies of neurotic depression[19] reviewed earlier, both classes of characterological deviations can develop serious affective illness, including melancholia.

Having reviewed the issue of chronology, one should next address *the length of time for which the mood disturbance has been present.* Is the change in mood isolated or transient, or is it associated with other complaints and signs that make it part of a syndrome? Have those complaints been present for several weeks (perhaps with a fairly clear-cut onset), or do they appear to have existed chronically with a recent exacerbation? A decision on this issue will help differentiate between episodic major affective illness and the other specific affective syndromes, notably dysthymia and cyclothymia.

Let us presume that the illness is episodic. Then the next important consideration in our decision tree becomes *the polarity of the illness.* Have there been both depressive and hypomanic/manic episodes in the past, or only one kind? To determine whether the disorder is bipolar or not is an extremely important clinical diagnostic decision, since bipolarity usually dictates a trial of lithium prophylaxis. Major bipolar illness, termed by some Bipolar I, refers to those individuals who meet the criteria for mania (see table 3.2). Usually such persons will also experience major depressive episodes. For the very small group of individuals who have recurrent mania dissociated from depression (nonbipolar mania), the illness is frequently secondary to other disease or its treatment.

There are also many subtle forms of bipolarity to be recognized. Categorized under "other specific disorders," these range from cyclothymia, in which intermittent and subsyndromic swings occur irregularly, often presenting as behavioral, interpersonal, marital, vocational, or substance abuse problems (Table 3.6) to recurrent syndromal depressions with a hypomanic response which occurs when the individual is treated with tricyclic antidepressants or electroconvulsive therapy. Some recurrently depressive individuals experience ego-syntonic and pleasant highs during which useful or creative endeavors may be pursued. Patients in this last group usually do not complain of elevated moods and therefore will not give such a history unless expertly questioned; bipolarity is sometimes suggested by relatives or associates who are more aware of the intermittent hypomanic drive. Pharmacologically occasioned hypomanic episodes may also be observed by the clinician during long-term care.[36] In all three of these subtle, sometimes termed Bipolar II, forms of affective disorder, the family

Table 3.6 Manifestations of Cyclothymia*[a]*

A. General

1. Onset in teens or early adulthood
2. Clinical presentation as a personality disorder (patient often unaware of "moods" per se)
3. Short cycles—usually days—which are recurrent in an irregular fashion, with infrequent euthymic periods
4. May not attain full syndrome for depression and hypomania during any one cycle, but entire range of affective manifestations occurs at various times
5. "Endogenous" mood changes, i.e., often wake up with mood

B. Biphasic course

1. Hypersomnia alternating with decreased need for sleep (although intermittent insomnia can also occur)
2. Shaky self-esteem which alternates between lack of self-confidence and naive or grandiose overconfidence
3. Periods of mental confusion and apathy, alternating with periods of sharpened and creative thinking
4. Marked unevenness in quantity and quality of productivity, often associated with unusual working hours
5. Uninhibited people-seeking (that may lead to hypersexuality) alternating with introverted self-absorption

C. Behavioral manifestations

1. Irritable–angry–explosive outbursts that alienate loved ones
2. Episodic promiscuity, repeated conjugal or romantic failure
3. Frequent shift in line of work, study, interest, or future plans
4. Resort to alcohol and drug abuse as a means for self-treatment or augmenting excitement
5. Intermittent financial extravagance

---

*[a]*From reference 43.

history is often positive for manic illness, providing a diagnostic clue to the bipolar nature of the disorder.

For the majority of individuals with major illness, the episodes will be of depression alone. Then, when examining the individual, it is important to decide whether the syndrome has become autonomous or *melancholic*.

It was Gillespie in 1929 in Edinburgh who first suggested the concept of autonomy as a valuable clinical distinction to be made in serious depressive episodes.[44] He noted that regardless of whether or not the illness appeared to be precipitated by stress, once established, autonomous depressions pursued their clinical course without being influenced to any significant extent by psychosocial factors introduced into the hospital environment. Such depressions he distinguished from "reactive" cases, in which psychosocial intervention by staff or relatives was a potent force in alleviating or aggravating the mood state. Hospitalization itself in such instances often was sufficient to induce apparent remission. This nosological scheme, which aids in predicting treatment response, has been incorporated into *DSM-III* using the term *melancholia* to identify the development of autonomy. Conceptually one may imagine that the disorders, regardless of etiology, have entered a final common path of pathophysiological dysfunction resistant to environmental perturbation.[45] The presence of melancholia (Table 3.7, see page 60) is usually a clear indication that psychotherapy is not the initial treatment of choice. Tricyclic pharmacotherapy or electroconvulsive therapy is ordinarily required to lyse the episode and render the patient responsive to meaningful interpersonal discourse.

Table 3.7 *DSM-III* Criteria for Melancholia

Melancholia. Loss of pleasure in all or almost all activities, lack of reactivity to usually pleasurable stimuli (doesn't feel much better, even temporarily, when something good happens), and at least three of the following:
1. Distinct quality of depressed mood, i.e., the depressed mood is perceived as distinctly different from the kind of feeling experienced following the death of a loved one
2. The depression is regularly worse in the morning
3. Early morning awakening (at least two hours before usual time of awakening)
4. Marked psychomotor retardation or agitation
5. Significant anorexia or weight loss
6. Excessive or inappropriate guilt

An additional qualification which may be chosen when using the *DSM-III* classification to describe both major depression and mania is *the presence of psychotic features*. These may take the form of affective delusions or hallucinations that are understandable in the context of the mood change, such as delusions of inferiority or grandiosity, or mood-incongruous delusions or hallucinations, such as Schneiderian symptoms that have now been shown to occur at the nadir of depression or the height of manic despair. These mood-incongruous features, originally postulated by Kraepelin to occur in affective illness, have been replicated by Clayton and colleagues,[46] Carlson and Goodwin,[47] and Abrams, Taylor, and Gaztanaga,[48,49] and have been summarized by Pope and Lipinski.[50] The basis for including these psychotic affective states—even with mood-incongruous features—in the core group of affective rather than schizophrenic disorders is that familial patterns, treatment response, and outcome parameters are indistinguishable in these patients from those with nonpsychotic forms of major affective disorder.[43]

Chronic unremitting depressions, termed dysthymia in *DSM-III*, represent a major challenge to nosologists. While chronic loss of interest or pleasure, frequently coupled with depressed mood, is a central feature, there are many other mild somatic and behavioral factors that occur and appear to be specific to the suffering individual. Many of these chronically disturbed persons have in the past been called neurotic, and the outlook for remission of the disturbance is poor. In our opinion, it is unlikely that this cluster is homogeneous other than sharing this poor prognosis, and a much more detailed follow-up of such persons, who are frequent visitors to the general physician, is required. This is an area in which careful use of the two remaining axes (IV and V) of *DSM-III*, those devoted to the severity of psychosocial stress and evaluation of adaptive social function, can contribute valuable guidance to future clinical research.

Outcome Studies of the Diagnostic Criteria

It is only through validation, through matching the phenomenological clustering observed with family history patterns, biological variables, natural clinical course, and responses to treatment, that the accuracy of the nosology now adopted will be determined. If an individual clinician faithfully follows an organized and broadly agreed-upon classification such as that proposed in the new diagnostic manual, the limitations and the advantages that grow from an arbitrary system will be quickly recognized. One major positive feature is that a common language will develop and the confusing or incorrect elements of the nosology will be more rapidly recognized.

National collaborative research is an organized extension of such individual diligence. As such studies grow in volume and our knowledge of basic pathophysiological mechanisms improves, we can look forward to a continued evolution of our present nosology.

References

1. Lewis, A. States of depression: Their clinical and aetiological differentiation. *British Medical Journal,* 1938, *2,* 875–878.
2. Irvine, W. *Apes, angels and Victorians; Darwin, Huxley and evolution.* New York: McGraw-Hill, 1955.
3. Milton, J. Areopagitica. Cited in the *The Oxford dictionary of quotations.* London: Oxford University Press, 1953.
4. *DSM-III: Diagnostic and statistical manual of mental disorders,* ed. 3. Washington, D. C.: American Psychiatric Association, 1980.
5. Kiloh, L. G., Andrews, G., Neilson, M., and Bianchi, G. N. The relationship of the syndromes called endogenous and neurotic depression. *British Journal of Psychiatry,* 1972, *121,* 183–196.
6. Kendell, R. E. The classification of depressions: A review of contemporary confusion. *British Journal of Psychiatry,* 1976, *129,* 15–28.
7. Robins, E. Categories versus dimensions in psychiatric classification. *Psychiatric Annals,* 1976 *6,* 39–55.
8. Roth, M. and Barnes, T. R. E. The classification of affective disorders: A synthesis of old and new concepts. *Comprehensive Psychiatry,* 1981, *22,* 54–77.
9. Kraepelin, E. *Manic–depressive insanity and paranoia.* Edinburgh: E. and S. Livingstone, 1921.
10. Thomson, K. C., and Hendric, H. S. Environmental stress in primary depressive illness. *Archives of General Psychiatry,* 1972, *26,* 130–132.
11. Leff, M. J., Roatch, J. F., and Bunney, W. E. Environmental factors preceding the onset of severe depressions. *Psychiatry,* 1970, *33,* 293–311.
12. Hirschfeld, R. M. Situational depression: Validity of the concept. *British Journal of Psychiatry,* 1981, *139,* 297–305.
13. Robins, E., and Guze, S. B. Classification of affective disorders: The primary–secondary, the endogenous–reactive, and the neurotic–psychotic concepts. In T. A. Williams, D. M. Katz, and J. A. Shield, (Eds.), *Recent advances in the psychobiology of the depressive illnesses.* Washington, D. C.: U. S. Government Printing Office, 1972, pp. 283–292.
14. Feighner, J. P., Robins, E., Guze, S. B., Woodruff, R. A., Winokur, G., and Munoz, R. Diagnostic criteria for use in psychiatric research. *Archives of General Psychiatry,* 1972, *26,* 57–63.
15. Goodwin, D., and Guze, S. B. *Psychiatric diagnosis.* New York: Oxford University Press, 1979.
16. Murphy, G. E., Woodruff, R. A., Herjanic, M., and Fischer, J. R. Validity of the diagnosis of primary affective illness: A prospective study with a five-year follow-up. *Archives of General Psychiatry,* 1974, *30,* 751–756.
17. Woodruff, R. A., Murphy, G. A., and Herjanic, M. The natural history of affective disorders. I. Symptoms of 72 patients at the time of index hospital admission. *Journal of Psychiatric Research,* 1967, *5,* 255–263.

18. Clayton, P. J., and Lewis, C. E. The significance of secondary depression. *Journal of Affective Disorders*, 1981, *3*, 24–35.

19. Akiskal, H. S., Bitar, A. H., Puzantian, V. R., Rosenthal, T. L., and Walker, P. W. The nosological status of neurotic depression: A prospective 3–4 year follow-up examination in the light of the primary–secondary and the unipolar–bipolar dichotomies. *Archives of General Psychiatry*, 1978, *35*, 756–766.

20. Akiskal, H. S., Rosenthal, R. H., Rosenthal, T. L., Kashgarian, M., Khani, M. K., and Puzantian, V. R. Differentiation of primary affective illness from situational, symptomatic and secondary depressions. *Archives of General Psychiatry*, 1979, *36*, 635–643.

21. Kupfer, D. J., and Thase, M. E. The use of sleep laboratory in the diagnosis of affective disorders. *Psychiatric Clinics of North America*, 1983, *6*, 3–25.

22. Akiskal, H. S., Lemmi, H., Yerevanian, B., King, D., and Belluomini, J. The utility of the REM latency test in psychiatric diagnosis: A study of 81 depressed outpatients. *Psychiatric Research*, 1982, *7*, 101–110.

23. Akiskal, H. S., Djenderedjian, A. H., Rosenthal, R. H., and Khani, M. K. Cyclothymic disorder: Validating criteria for inclusion in the bipolar affective group. *American Journal of Psychiatry*, 1977, *134*, 1227–1233.

24. Keeler, M. H., Taylor, C. I., and Noonan, D. L. Are all recently detoxified alcoholics depressed? *American Journal of Psychiatry*, 1979, *136*, 586–588.

25. Dunner, D. L., Fleiss, J. L. Addonizio, G., and Fieve, R. R. Assortative mating in primary affective disorder. *Biological Psychiatry*, 1976, *11*, 43–51.

26. Leonhard, K. *The classification of endogenous psychoses*, 5th ed. New York: Irvington Publishers, 1979.

27. Angst, J. (Ed.). *Classification and prediction of outcome of depression*. Stuttgart: F. K. Schattauer Verlag, 1974.

28. Perris, C. A study of bipolar (manic–depressive) and unipolar recurrent depressive psychoses. *Acta Psychiatrica Scandinavica*, 1966, *42*, (suppl. 194), 7–189.

29. Winokur, G., Clayton, P. J., and Reich, T. *Manic–depressive illness*. St. Louis: C. V. Mosby, 1969.

30. Spitzer, R. L., Endicott, J., and Robins, E. Research diagnostic criteria: Rationale and reliability. *Archives of General Psychiatry*, 1978, *35*, 773–782.

31. Goodwin, F., and Bunney, W. E., Jr. A psychobiological approach to affective illness. *Psychiatric Annals*, 1973, *3*, 19–53.

32. Fieve, R. R., and Dunner, D. L. Unipolar and bipolar affective states. In F. Flach and S. Draghi (Eds.), *The nature and treatment of depression*. New York: John Wiley and Sons, 1975, pp. 145–160.

33. Depue, R. A., and Monroe, S. M. The unipolar–bipolar distinction in the depressive disorders. *Psychological Bulletin*, 1978, *85*, 1001–1029.

34. Gershon, E. S. The search for genetic markers in affective disorders. In M. A. Lipton, A. DiMasgo, and K. F. Killam (Eds.), *Psychopharmacology: A generation of progress*. New York: Raven Press, 1978, pp. 1197–1212.

35. Akiskal, H. S. The bipolar spectrum: New concepts in classification and diagnosis. In L. Grinspoon (Ed.), *Psychiatry update: The American Psychiatric Association annual review*, vol. II. Washington, D. C.: The American Psychiatric Association Press, 1983, pp. 271–292.

36. Akiskal, H. S., Walker, P. W., Puzantian, V. R., King, D., Rosenthal, T. L., and Dranon, M. Bipolar outcome in the course of depressive illness: Phenomenologic, familial and pharmacologic predictors. *Journal of Affective Disorders*, 1983, *5*, 115–128.

37. Gershon, E. S., Mark, A., Cohen, N., Belizon, N., Baron, M., and Knobe, K. E. Transmitted factors in the morbid risk of affective illness: A controlled study. *Journal of Psychiatric Research*, 1975, *12*, 283–299.

38. Depue, R. A., Slater, J. F., Wolfstetter-Kausch, H., Klein, D., Goplerud, E., and Farr, D. A behavioral paradigm for identifying persons at risk for bipolar depressive disorder: A conceptual framework and five validation studies. *Journal of Abnormal Psychology* (monograph), 1983, *90*, 381–437.

39. Akiskal, H. S. Dysthymic disorder: Psychopathology of proposed chronic depressive subtypes. *American Journal of Psychiatry*, 1983, *140*, 11–20.

40. Paykel, E. S. Depressive typologies and response to amitriptyline. *British Journal of Psychiatry*, 1972, *120*, 147–156.

41. Pilowsky, I., Levine, S. and Boulton, D. M. The classification of depression by numeral taxonomy. *British Journal of Psychiatry*, 1969, *115*, 937–945.

42. Klerman, G. L., Endicott, J., Spitzer, R., and Hirschfeld, R. M. A. Neurotic depressions: A systematic analysis of multiple criteria and meanings. *American Journal of Psychiatry*, 1979, *136*, 57–61.

43. Akiskal, H. S. Affective disorders: Special clinical forms. In H. S. Akiskal (Ed.), *The psychiatric clinics of North America*, vol. 2, no. 3. Philadelphia: W. B. Saunders, 1979.

44. Gillespie, R. D.. The clinical differentiation of types of depression. *Guy's Hospital Reports*, 1929, 79–83.

45. Whybrow, P. C., and Parlatore, A. Melancholia, a model in madness; A discussion of recent psychobiologic research into depressive illness. *Psychiatry in Medicine*, 1973, *4*, 351–377.

46. Clayton, P. J., Pitts, F. N., and Winokur, G. Affective disorder. IV. Mania. *Comprehensive Psychiatry*, 1965, *6*, 313–322.

47. Carlson, G., and Goodwin, F. The stages of mania. *Archives of General Psychiatry*, 1973, *28*, 221–288.

48. Abrams, R., Taylor, M. A., and Gaztanaga, P. Manic–depressive illness and paranoid schizophrenia. *Archives of General Psychiatry*, 1974, *31*, 640–642.

49. Taylor, M. A., Gaztanaga, P., and Abrams, R. Manic–depressive illness and acute schizophrenia: A clinical, family history, and treatment response study. *American Journal of Psychiatry*, 1974, *131*, 678–682.

50 Pope, H. G., Jr., and Lipinski, J. F., Jr., Diagnosis in schizophrenia and manic depressive illness. *Archives of General Psychiatry*, 1978, *35*, 811–828.

Elements of Present Knowledge

This second part of the book is divided into four chapters; three of them review information specifically relevant to our present knowledge of affective illness, but each does so from a different perspective (Chapters 4, 5, and 7). Chapter 6 offers an introduction to basic neurobiology for those readers unfamiliar with the subject.

The experience of attachment and loss is not unique to human beings; many of the higher mammals demonstrate such behavior, and the phenomenon has been extensively studied in primates. We review these studies in Chapter 4 and extend their significance to the human experience of grief and mourning. The evidence suggests that attachment to others is a primary drive, as are sexual activity and eating. Disturbance of this process in infancy may jeopardize normal attachment in later life, leading to psychopathology and perhaps predisposing an individual to depressive illness. The evidence for such a sensitization is reviewed.

In Chapter 5 we take up the importance of symbolism and abstraction in mental life. These functions appear to be a unique capacity of the human brain, adding a dimension of extraordinary complexity to psychopathology. The seminal contributions made by Freud and the theories of psychoanalysis to our dynamic understanding of mental life are reviewed from a historical perspective. Freud's own experiences within his family and his culture are used as background to the development of his dynamic constructs of intrapsychic phenomena and unconscious drives, especially as they are reflected in his differentiation of mourning from melancholia. Later Freudians expanded these concepts to include other intrapsychic processes: narcissism and dependence (Rado), primary infant/parent attachment (Gero), infantile aggression/helplessness (Klein), self-esteem and the discrepancy between the preferred and perceived image of self (Bibring, Jacobson). It becomes apparent that the theme of attachment and loss runs through all those theories, providing resonance with the primate and infant studies discussed in Chapter 4.

Chapter 6 offers, initially through illustrations from endocrinology, a review of some basic principles and factual information pertinent to modern neurobiology. Because of its importance in understanding many of the studies and theories that underpin biological research in the mood disorder, a specific outline of the neurochemistry of biogenic amines is provided. A brief diversion exploring how neurochemical and neuroendocrine events reflect and interact with the challenges of the extended environment completes the chapter.

A synopsis of the biological research that is pertinent to affective disorder is Provded in Chapter 7. From the serendipitous beginnings of modern psychopharmacology, we review the current evidence for disturbance of biogenic amine metabolism in depression and mania. The complementary investigations in neurophysiology and endocrinology are also discussed, and those areas of research are highlighted in which consistent findings of biological dysfunction have emerged.

Attachment and Loss

"Charlotte," said Wilbur. "We're all going home today. The Fair is almost over. Won't it be wonderful to be back home in the barn cellar again with the sheep and the geese? Aren't you anxious to get home?"

For a moment Charlotte said nothing. Then she spoke in a voice so low Wilbur could hardly hear the words.

"I will not be going back to the barn," she said.

Wilbur leapt to his feet. "Not going back?" he cried. "Charlotte, what are you talking about?"

"I'm done for," she replied. "In a day or two I'll be dead. I haven't even strength enough to climb down into the crate. I doubt if I have enough silk in my spinnerets to lower me to the ground."

Hearing this, Wilbur threw himself down in an agony of pain and sorrow. Great sobs racked his body. He heaved and grunted with desolation. "Charlotte," he moaned. "Charlotte! My true friend!"

E. B. White[1]
Charlotte's Web (1952)

The Experience of Grief

For those of us who have grieved, the experience is unforgettable. After initial numbness and disbelief, comes an utter and profound sadness. The chest feels heavy and the appetite is lost, seemingly forever. Then there arises a restlessness, an urge to pace, an inability to concentrate. Insomnia drains the energy, and crying spells occur. One becomes irritable, and while seeking support from friends is sometimes comforting, it is only with a supreme effort that one can involve oneself in their lives. This acute stage gradually subsides, but grief does not disappear. Special words, memories, places can reactivate the agonized feelings and preoccupations weeks, months, or even years later. Nonetheless, over time, as the lost person is discussed, important memories are shared with others and mixed feelings are explored, the acute impact of the grief lessens and one is able to return to daily routines.

Wilbur, the anthropomorphic pig in *Charlotte's Web,* E. B. White's classical tale of barnyard joys and struggles, dealt with his grief in an unusual way. Faced with the impending death of the spider Charlotte, who was his close friend and rescuer, he bribed Templeton the rat to save her progeny. Although most of these young spiders left Wilbur when they were old enough to be independent—provoking further grief—three of Charlotte's daughters stayed. The important bond with Charlotte had been preserved through her children.

Charlotte's Web is a compelling and timeless story of commitment and attachment. The sadness which children share with Wilbur as he learns that Charlotte will soon die is an empathic awareness of the great fondness Wilbur has for his spider friend. We are all reminded of our own relationships, for without closeness to others, to ideals, and to places, a sense of loss and subsequent grieving cannot exist. As noted by Parkes, the British authority on bereavement and loss, grief is the price of commitment.[2]

We should not dismiss these changing relationships of the barnyard as anthropomorphic nonsense, for as animals, we are not alone in forming bonds both with those of our own species and with others. The family dog, cat, horses, sheep, and even cows exhibit behaviors of loneliness and sadness when separated from us and from their own kind. Konrad Lorenz,[3] in *Man Meets Dog,* describes with considerable poignancy the events surrounding his separation from a very special dog named Stasi.

> Stasi was born in our house in early spring, 1940, and was seven months old when I adopted her as my own, and began to train her.
>
> After two short months, my bond with this dog was broken by the force of destiny: I was called to the University of Konigsberg as professor of psychology and I left my family, home and dogs on September 2, 1940. When I returned at Christmas for a short holiday, Stasi greeted me in a frenzy of joy, demonstrating that her great love for me was unchanged. She could do everything I had taught her, just as well as before, and was indeed exactly the same dog as I had left behind me four months previously. But tragic scenes were enacted when I began to prepare for my departure. Many dog-lovers will know what I mean. Even before the suitcase packing—the visible sign of departure—had started, the dog became noticeably depressed and refused to leave my side for an instant. With nervous haste, she sprang up and followed every time I left the room, even accompanying me to the bathroom. When the trunks were packed and my departure became imminent, the misery of poor Stasi waxed to the point of desperation, almost to a neurosis. She would not eat and her breathing became abnormal, very shallow and punctuated now and then by great, deep sighs. Before I left, we decided to shut her up, to prevent her making a violent attempt to follow me. But now, strangely, the little bitch, who had not left my side for days, retired to the garden and would not come when I called her. The most obedient of all dogs had become refractory, and all our efforts to catch her were in vain.

What is true of the dog becomes even more obvious in the primate, and we have learned much about ourselves from a study of the attachment behavior of these animals. It would seem that close social bonds are an important component of psychobiological stability for many higher animals. Furthermore, disruption of these bonds can be expected to produce a disorganizing perturbation, which we label subjectively as grief, and which to the observer appears as behavioral withdrawal and regression. Therefore, not surprisingly, many clinicians and theorists have been interested in the relationship of loss and grieving to the development of depression.

The pain of grief is a central theme in our literature and in religious experience. Robert Burton discussed at length the origins and treatment of "love melancholy."[4] Freud, in *Mourning and Melancholia*, described in detail the experience of mourning and compared and contrasted it with melancholia.[5]

In this chapter we review studies of attachment in human beings and in primates, including the acute grief reaction that occurs when this attachment is broken. We shall confine ourselves to the behavioral changes that such disruptions produce. Discussion of the intrapsychic dimension, a crucial element in the understanding of human mourning and subjective experience of depressive illness, is taken up in Chapter 5.

Bereavement and Its Sequelae

The fire in the Coconut Grove Restaurant of Boston was a major tragedy of 1942. Many were trapped and died in the burning building, and many of those who escaped did so without their family or friends.

Eric Lindemann, a Boston psychiatrist, studied those bereaved by the fire and also their close relatives. He compared them with three other groups of individuals: relatives of members of the armed forces who had been killed in the battles of the Second World War, families of individuals who died in the hospital, and a group of patients in psychotherapy for neurosis who had lost a relative during the period of treatment.

Data were gathered through a detailed psychiatric history and interviews. While there was no systematic recording of symptoms, Lindemann's work was the first of a series of studies that attempted to outline the syndrome of normal grief. Lindemann described five essential features:

1. Somatic distress, such as tightness in the throat, shortness of breath, sighing respirations, an empty feeling in the abdomen—all general discomforts, described as coming in waves and usually but not always environmentally stimulated
2. A preoccupation with the image of the dead person. Frequently, when off guard, others would be transiently misidentified as being the dead person. Even transient auditory and visual hallucinosis of the lost person was reported in some cases
3. Guilt surrounding the death and the mourner's contribution to it
4. Angry and hostile reactions
5. Loss of the usual daily patterns of routine conduct

This clinical study has become a classic in the psychiatric literature, but the special nature of the samples studied limited generalization and necessitated further study.[6]

Marris conducted a retrospective survey in London of 72 young women who had been widowed. The interviews, which were unstructured, were conducted about 2 years after the husband's death. The most frequent reactions described by the women were a lasting deterioration in health, continued difficulty in sleeping, withdrawal from social activity, and a loss of contact with the real world, coupled with a diffuse sense of anger.[7] Again, however, it is dangerous to generalize from these findings to the population at large, because of the nonsystematic review and the long period of time elapsed between the actual death of the spouse and the time of interview.

The most comprehensive and controlled series of studies to date have been those by Paula Clayton and associates from Washington University in St. Louis. In one study these authors examined the morbidity and mortality of widowhood in a series of systematic interviews of 109 bereaved women. The most common manifestations, displayed by over 50% of the widows, were depressed mood, disturbance of sleep, and crying spells. These particular experiences clustered in the first few months of the bereavement, and although the psychological disturbances appeared in many instances to be indistinguishable from those of depressive illness, they were accepted by the individual and the relatives as normal. These individuals who went through "normal" bereavement did *not* show any increase, when compared to the general population, in visits to hospital or physicians, in treatment for depression, or in the use of psychotropic drugs.[8]

This, of course, is in stark contrast to patients with primary affective illness who experience their condition as alien—or discontinuous with their usual self—and frequently seek professional assistance. Another major and conspicuous difference between mourners and depressed persons is that those who grieve do not experience psychomotor retardation, fears of going insane, disturbances of self-regard, ideas of self-condemnation, or suicidal preoccupation. Such characteristics are common in those clinically depressed.

The prospective studies of the St. Louis group suggest that the majority of persons who are bereaved cope with the process without developing psychiatric illness. Interestingly, however, it was found that younger people who lost a spouse showed more of the somatic depressive symptoms than matched married controls and older widows and widowers. One may speculate about these differences. Is there in youth a lack of preparation for death, of a sense of injustice because of the early loss of the partner? Further research will be required to clarify these propositions.

What of the small group of individuals, 2%–5% of a bereaved population, who do not recover spontaneously from the process of grief and mourning and go on to develop clinical depression? It is probable that a confluence of factors determines this outcome:

1. The nature of the relationship with the individual who is lost
2. The availability of supportive family members and friends
3. The physical state of the bereaved at the time of loss
4. Familial or genetic predisposition to depressive illness
5. The previous experience adapting to the disorganizing effects of separation, and whether this was successfully accomplished

Precipitation or exacerbation of somatic illness may also occur in the postbereavement phase. Parkes[2] has shown that in bereaved individuals predisposed to coronary artery disease, the 6 months following bereavement represent a special risk period for myocardial infarction. Grief can thus literally "break the heart." This outcome, although rare, is probably accounted for by the adverse cardiovascular impact of increased sympathetic nervous system activity. In general surveys Clayton and associates[8] have shown that the mortality in the first year of bereavement is no different from that in matched controls. It appears, therefore, that loss may constitute an overwhelming challenge which can result in death, but only for those in whom a predisposition to coronary heart disease exists.

Similarly, it is reasonable to suppose that bereavement does not *cause* clinical depression, but may determine the *timing* of such depression in a predisposed individual. However, the Clayton et al. study failed to document a significantly increased incidence of familial affective illness—taken as an indirect measure of genetic predisposition—in those few bereaved individuals who progressed from grief to clinical depression. This suggests that in the small proportion of individuals in whom bereavement does lead to depressive illness, there is selective interaction with other psychosocial factors. Such a factor is the absence of supportive family members, who may act as emotional buffers and gradually neutralize the pain of the loss. The potential of bereavement to precipitate depression *in the absence* of social support has been demonstrated in a study conducted in Boston by Maddison and Walker,[9] and later replicated in Australia.[10] Whether childhood experience of loss is also a factor that modifies the response to adult bereavement is considered later in this chapter.

Attachment and Its Formation

If grief, which we know to be a potent disorganizing force, grows from disruption of attachment, then it follows that we may learn something of the genesis of grief and other separation reactions, including the precipitation of clinical depression, by first exploring the nature of attachment behavior itself.

Let us return to Wilbur the pig as an example of attachment formation. As the runt of the litter he had been saved from the axe by Fern, the farmer's daughter, who loved the little pig as if he were her own child. She built his house, fed him, stroked him, and watched him play. Long before Wilbur met Charlotte the spider, Fern had protected and nurtured him. Thus, Fern was a stable and reliable part of Wilbur's early life.

When Wilbur was moved to the novel surroundings of another farm, Fern continued to visit every afternoon. The bond was a stable one, a consistent parameter promoting a stable psychological state with minimal adaptive demand (cf. Chapters 8 and 9). Within the security of this bond Wilbur learned to adapt to the harsh realities of the farmyard. In adolescence he briefly toyed with running away, but decided he preferred his bondage and his secure and trusting attachments. In time Wilbur grew to be a shrewd and capable pig, but without that special early bond with Fern, as every reader of *Charlotte's Web* knows, the outcome might have been very different.

How do such attachments develop and why? What are the effects if the opportunity is never afforded? Is it their disruption that leads to grief?

Primate studies at the University of Wisconsin have been very helpful in answering some of these questions. Working with rhesus monkeys, the Harlows have identified five major affectional systems or "love bonds," which are probably similar to those found in humans. They fall into several general categories: the infant–mother bond, mother–infant bond, peer–peer bond, father–infant affectional system, and heterosexual bonding.[11–13] Much has been written about how each of these systems develops. John Bowlby's recent volumes on attachment and separation behavior—in which he interweaves field studies on animal behavior and primate laboratory research with clinical findings—are considered the major reference on the subject.[14-16]

Initially, the attachment of the infant rhesus monkey to its mother is primarily reflex; only gradually does it enter the stage of attachment and comfort where voluntary responses replace the automatic ones (a stage thus called "voluntary attachment"). Physical contact is a primary factor in this early bonding. In the next stage the infant grows to feel secure in the mother's presence and eventually secure enough to broaden its social contacts and to explore the extended environment. Human infants go through similar processes in developing reciprocal relationships with their mother, albeit more slowly.

A number of studies in rhesus monkeys also have documented the importance of early peer experience. Affection builds and develops not only between infant and mother but between members of a group, a process that appears critical for subsequent healthy adaptation. In human studies we have for too long ignored the importance of such peer relationships. Their disruption is probably one of the contributory factors to depression in adolescence and, hence, potentially an important factor in determining adaptation in the adult years.

These and other affectional bonds have played a key role in primate evolution. As the Harlows have noted, "without group living, primates would have had to develop specialized individual protection as, for example, have the cats and they could not have afforded the luxury of a long developmental period prior to and after birth."[13] Obviously this is even more true of human beings.

Attachment as a Primary Drive

A prevalent explanation of attachment behavior is known as the "secondary drive" theory. Basically, this theory maintains that the infant has a number of physiological needs which must be met and attachment to the mother develops naturally as these are attended. Indeed, a number of clinical tenets in psychoanalytic teaching have been based on such a concept, particularly the hypothetical construct of the "oral phase."

This secondary drive thesis was initially questioned by the Nobel prize-winning work of Lorenz,[17] who demonstrated that attachment behavior of ducklings and goslings in the wild could develop without their receiving any nourishment. These birds developed attachment bonds (which Lorenz called imprinting) when exposed to other birds or Lorenz himself at certain critical times following birth. These experiments have been repeated, basically replicated, and refined in other species.

The laboratory work of the Harlows has further demonstrated that mother–infant attachment is influenced by a variety of variables other than physiological drive reduction. Foremost among these is contact comfort. Infant rhesus monkeys prefer a nonfeeding cloth mother to a wire surrogate that provides food. Of course, if contact comfort is equally available from two surrogates, the majority of infants will choose the feeding mother. A number of other variables, including those of facial expression, motion, and temperature, are also important. These factors appear to reinforce the primary drive and ensure the proper development of the attachment bond.[13]

There is thus persuasive evidence that attachment behavior is a primary drive which, while it needs reinforcement by the behavior of the mother, is not subservient to other drives for food or satisfaction of oral–libidinal urges. Innate propensities of primates, including humans, foster the development of these close social bonds. However, a successful primary attachment experience with a parent or parents is essential to adequate development of the capacity for later social bonding. Where such needs are inadequately met, there may be distortion in subsequent relationships and extreme sensitivity to disruptions of interpersonal attachment.

Failure to Develop Primary Attachment

There are situations in both human and nonhuman primates where the development of primary attachment bonds is impaired or prevented. This may occur when a human infant is socially isolated from birth, and may be experimentally induced in other primates by isolation rearing. We will briefly review both human and animal studies. Interestingly, the findings are similar.

In primates two basic kinds of social isolation experiments have been conducted: partial and total. In partial social isolation, monkeys are reared from birth in wire cages where they can see other animals in the room but are allowed no physical contact with them. Total social isolation involves rearing from birth in an isolation chamber which allows no visual or tactile contact with other monkeys. Rhesus monkeys reared in such circumstances for the first 12 months of life and then removed and placed with other animals, have severe behavioral deficits. They spend most of their time huddled alone in a corner, rocking, self-clasping, and refusing to enter into peer play or other age-specific social encounters. They often engage in stereotypic behaviors, may mutilate themselves, and also exhibit unpredictable and inappropriate aggression. A common behavior has become known as the "floating limb phenomenon," in which the monkey spends long periods of time watching his arm float slowly by in front of him. When such animals reach puberty, appropriate sexual behaviors are virtually absent, and the females, if they ever become pregnant, are unsatisfactory mothers, especially with their first infant.

There are many critical variables in determining the effects of early social isolation, including the degree of isolation, the age at which it begins, and its duration. Not surprisingly, the most severe effects are produced when the isolation is total, begins at birth, and lasts for 6 months or more.

It is clear from this work with primates that failure of primary social attachment has devastating behavioral consequences over a long term. Moreover, most of the deficiencies are difficult to change.

Analogous situations have been reported among human beings. Although the totally socially isolated human infant is fortunately rare, lesser degrees of deprivation still have a destructive impact on normal psychobiological development. The social input to infants and children is variable in quality and consistency, depending on familial interaction patterns. Simple lack of physical contact with a caring adult is not uncommon, although difficult to ascertain in retrospect.

Some dramatic extremes, however, have been reported and are of heuristic importance here. One such is the report by Freedman of Albert and Anne. The mother of these children was said to have had a delusion that the children were defective and should therefore be isolated. Albert was reared from birth to 4 years in an 8- by 10-foot room with the only window covered with burlap. The mother entered the room only to diaper him and hand him his bottle of gruel. The results of this massive social deprivation are frighteningly similar to those seen in the rhesus monkey. When Albert was finally discovered, he exhibited severely retarded physical development and a lack of purposeful motor activity. He was incontinent, had no articulate speech, and would sit immobile for long hours, engaging in frequent head-banging and rocking behaviors. He was indifferent to other humans and did not seek to be held or comforted. He would eat almost anything and showed no evidence of complex social play. Although his physical development improved markedly during a course of hospitalization and subsequent foster home placement, he still showed no evidence of developing primary attachment behavior. For example, he would not mold his body to the person holding him, and did not respond differentially to approaches by people familiar to him and strangers.

Anne, Albert's sister, was confined to a separate room in the home for the first 6 years of life, except for a few visits to a physician. She showed similar physical symptoms and a lack of social development when hospitalized and later placed in a foster home.

The striking feature of both these children was that the period of massive early environmental deprivation resulted in severe deficits in motor, cognitive, and social skills. While these children improved substantially, there remained in both an inability to develop primary attachment bonds with others.[18]

Freedman contrasted the behavior of these two children with that of the "wolf children" who, having spent their early years running with a wolf pack, had a rich supply of social and physical stimulation, albeit somewhat atypical in form! There is anecdotal evidence to suggest that the wolf children did develop attachment behavior to each other and showed a grief response when one died.[18]

Perhaps it is, then, that primary attachment is a prerequisite of depression, and further, that there is a critical period for such attachment to develop in ourselves, just as in the fowl studied by Lorenz. The case of Albert and Anne and the primate studies suggest, for example, that early disruption of the primary drive may make later attachment impossible. In the absence of such attachment, the resulting psychopathology appears to be far more pervasive and destructive than that commonly found in affective illness.

Behavior following Separation in Human and Monkey Infants

Attachment to others appears to be a necessary component in the development of normal adaptive behavior in higher animals and man. In fact, a close bond is so important that once primary attachment has begun to develop, disruption brings outrage and protest from the young of such species.

Such a syndrome was first described in human infants in 1947 by the London analyst Rene Spitz.[19] This response to maternal separation, termed "anaclitic depression," has been confirmed and subdivided into distinct stages by the Robertsons[20] and Bowlby.[16] An initial "protest" stage is characterized by much motor restlessness, with attempts to attract attention and presumably the return of the parent. There follows a "despair" stage, characterized by weeping and reduced activity, and finally a "detachment" with withdrawal from activity, people, and the general environment. Indeed, should the mother return during this detachment phase, rejection of her initial attempts to reestablish the relationship usually occurs.

In general, the separation studies that involve nonhuman primates have found a biphasic response to separation, with agitation followed by withdrawal. However, the occurrence and nature of this response are very much influenced by such parameters as age of the infant, the quality of the preseparation relationship, the species, the type of separation, and length of separation. Disruption of an ongoing mother–infant attachment bond is a reliable and powerful method for inducing separation reactions in nonhuman primates. Also, disruption of an ongoing peer relationship can lead to a biphasic protest–despair response, the nature of which has been shown to be influenced by a series of variables including the neurobiological status of the animal prior to and during the separation.

The manipulation of such separation paradigms has led to the successful development of a depression model in monkeys; the model has proved helpful to our understanding of depressive behavior. First, it illustrates the unique potency of separation events in disrupting the psychobiological equilibrium of the organism. Second, the essential similarity between the human and animal syndromes—especially with respect to both behavioral and somatic manifestations—suggests that mood disorders are not unique to human beings but behaviors that are released under appropriate environmental challenge in various higher vertebrates. Third, it marks a methodological advance, a convenient strategy, permitting rigorous study of the multiple variables that interact with separation to produce the behavioral syndrome.

Separation and the Precipitation of Mood Disorders

Separation is a potent force in disrupting the psychobiological homeostasis of human and rhesus infants in a depressive direction. We have seen that bereavement may produce a grief syndrome which mimics some aspects of clinical depression. Hence it comes as no surprise that one of the models of pathogenesis of mood disorder which we reviewed in Chapter 2 was that of object loss.

Researchers have long sought to define the relevance of separation events to the production of clinical depression. In the psychoanalytical literature, object loss refers to any event that is experienced as a loss, whether it be of a person, one's material possessions, status, social roles, or symbolic possessions such as an ideal or a goal. The intrapsychic meaning of such losses is considered in the next chapter. Here we confine our discussion to separation events, which by comparison are easy to define operationally and are thus open to objective study.

The research literature linking separation events to psychopathology represents a new strategy which forms part of what has been termed "life events research." The thrust of this approach is to demonstrate significant quantitative and temporal links between life events and the onset of illness. Despite considerable methodological problems, this literature offers convincing evidence that the onset of a variety of medical and psychiatric disorders is frequently preceded by certain types of life events. Events that lead to separation—bereavement, divorce, loss of a love relationship—top the list of the Holmes–Rahe life inventory,[21] the prototype of instruments used in such studies.

The London psychiatrist Eugene Paykel has classified life events, relevant to depressive illness, into various categories (see Table 4.1).[22] Paykel's own findings and those from other studies strongly support the role of "exit" events and "undesirable" events as of prime importance in the genesis of depression.[23,24] The studies indicate that two other kinds of events—so called "entrances" into the social scene and "desirable" events—are rarely implicated. "Success depressions," also known as "paradoxical depressions" associated with nominally "desirable" events, have been reported, however, even though they are rare. It is such depressions that emphasize the importance of understanding the meaning of the life change to the individual concerned, for a "positive" event such as promotion, in concealing the threat of increased responsibility, may actually have a negative connotation. The range of preceding events reported by depressives is large and not necessarily confined to separations or even losses. For example: increase in the number of arguments with one's spouse, marital separation, beginning a new type of work, change in work conditions, serious personal illness, death of an immediate family member, serious illness of family member, or family members leaving home have all been included.

A number of methodological barriers hinders work in this area. First, ethics obviously dictate that investigators must depend on naturally occurring events. Thus

Table 4.1 Classification of Life Events[a]

	Examples
I. "Exit" events from social scene	Separation from lover
	Deaths
	Children leave home
II. "Undesirable" events	Sexual difficulties
	Physical illness
	Demotion/bankruptcy
III. "Desirable" events	Buying a house
	Graduation
	Promotion
IV. "Entry" events into social scene	Marriage
	Planned pregnancy
	Birth/adoption of child

[a]From Paykel.[22]

most studies are retrospective and potentially distorted by inaccuracies in the recall of past events. The depressed person in particular may focus upon past tragedies, over-reporting and inflating the significance of aversive life events in comparison to pleasurable ones. Also, thinking may be so slowed that events are remembered only with difficulty, if at all. Such methodological problems are not insurmountable, however, and there have been a number of controlled studies of the relationship between separation and the precipitation of depression.[25–28]

The majority have reported data suggesting that separations and related events frequently precede and precipitate the disorder, but the relationship is not a simple one. Depressives report experiencing, on the average, three to four times more life changes per unit time than do nondepressed controls. It is possible that the concurrent presence of other stressful life events increases the pathogenic impact of separation. However, separation does not specifically cause depressive illness; other forms of psychopathology can and do follow separation. Neither is separation a *sufficient* cause of depression. Most individuals who suffer separation, 95% according to Paykel's study, do not develop depressive illness. Nor is separation a *necessary* antecedent of depression, since it does not precede all depressions. Indeed, separation and related events often result from the depression itself, for the clinically depressed individual may engage in a series of behaviors that leads to alienation of those around him and thus to separation. Finally, in those depressed subjects who lack social support, separation may precipitate hospitalization.

Whether loss can also precipitate mania is an unresolved question. Conceptually, psychoanalytic theory explains manic behavior as a mask for depression and denial of the pain of loss. Lindemann, in his classic studies of bereavement, mentioned hyperactivity as an atypical manifestation of grief. "Maniacal" responses to bereavement are accepted as clinical entities in France, and two British studies suggest that a significant increase in the number of life events, particularly separations, does precede manic episodes.[29,30]

Sensitization to Adult Depression by Loss in Childhood

The relationship of childhood loss to the development of adult depression is complex. A number of questions need to be considered. Is it only the death of the nurturing parent that predisposes the infant to later depression, or is separation sufficient? Also, what effect has the *quality* of that parental relationship prior to the loss?

Retrospective studies are numerous, but many have inadequate or no controls. The principal question to be answered is whether the incidence of parental death in a depressive population exceeds that of the general population. Carefully controlled studies suggest that the incidence of parental death before the age of 15 is higher in men and women with adult depression.[31] However, this finding does not seem to be specific for depression, although there is evidence that patients with severe or psychotic depressive illness are more likely to have experienced permanent loss in childhood. Similarly, patients who attempt suicide more frequently have lost a parent through death during their childhood. There are two age periods during which the death of a parent is statistically linked with depressive psychopathology in adult life: the loss of either the mother or the father during the first 5 years of life and, more strikingly, loss of the father between the ages of 10 and 14 years.

Regarding the second question, the studies that link nonpermanent separation in childhood (i.e., separation other than through parental death) to adult depression are fewer in number, and the findings less clear. At least one retrospective study,[31] however, has found that depressed patients, in comparison to nondepressed neurotic controls, more frequently experienced such separations. Several other studies have tried to trace the impact of divorce, separation, and other forms of family breakup on subsequent overall mental adaptation. The data are problematic. In general, such early-life experiences seem to predispose to increased incidence of broadly defined psychiatric disturbances including, but not limited to, depressive disorders.

While we may attempt with all due statistical rigor to correlate these retrospective data with depression in adult life, we are limited by our inability to gauge the *quality* of the sustaining factors in the social situation after the loss, and the individual resources of the child. These unique human perspectives—the components that provide the nuance of everyday adaptation and failure—probably cannot be fully evaluated even by the person who experiences them. Hence, the question of how the *quality of the parental relationship* impairs or enhances adaptation to loss must remain largely one of clinical inference and speculation. We shall return to this issue when considering the contribution of intrapsychic mechanisms to mood disorder.

Some Conclusions and Comment

Human beings are social animals. Attachment to other people is a primary drive, independent of food or sexual need, which usually leads to close bonding between ourselves and key people in our environment. Attachment to parents is the prototype of many similar bonds which are subsequently formed with siblings, peers, and lovers.

In primates and even in human beings there appears to be a critical time in infancy during which the primary attachments are explored and developed. Reinforcement of these infant behaviors by nurturing individuals is essential to normal psychobiological growth and development. Isolation studies clearly demonstrate that in primates the prevention of bond formation leads to grossly deviant behavior. Disruption of developing attachment bonds in primate and human infants leads to protest and later withdrawal. Hence a capricious or depriving environment can disorganize the normal psychobiological processes of attachment behavior.

Such a situation is well illustrated by the work of Engel and Reichsman with the infant Monica. These longitudinal investigations began in 1953 when Monica, aged 15 months, was admitted to hospital in a very withdrawn state. Because of esophageal atresia the infant had been fed since birth through a gastric fistula, but she had received little nurturing beyond that because of her mother's depressive illness. This had dominated the period from 6 to 11 months of Monica's life, when her mother was able to do little more than feed her on an essentially random schedule. During the depressed period the mother reported finding it particularly difficult to respond to her child's crying. In Engel's view the protest phase of the response to separation had become exhausted (by the age of 1 year Monica had given up crying) and a state of conservation–withdrawal had supervened. This passive sleeplike behavior, associated with virtually no motor activity and a cessation of gastric secretion, appeared to be an extension of the usual withdrawal phase seen in response to separation. It reflected, in Monica's case, a desperate adaptation to the depriving environment.[32]

The authors hypothesized that if prolonged this physiological state becomes associated during development with a subjective sense of helplessness. Later feelings of hopelessness may be added as the individual consciously recognizes the inadequacy of his or her ability to respond. This basic psychobiological process is in sharp contrast to the active–engaging posture which the growing infant adopts when secure in an environment that offers nurturing reinforcement.

It is not to be inferred, however, that all separation leads to pathology. Varying degrees of separation and loss are part of daily environmental challege. When graded within the ability of the infant, these events lead to successful psychobiological adaptation which builds autonomy and an objective regard for the self. During this process of growth, the feelings of helplessness and hopelessness have been described by Engel as "signal functions"—warnings which permit anticipation of special traumas that have been experienced previously and now again threaten to overwhelm the individual. If successful adaption does not occur, then the psychobiological process of withdrawal is precipitated.

These concepts are very helpful in understanding the mechanisms through which incomplete resolution of the attachment process in infancy and development can jeopardize subsequent bond formation in adult life. A particular vulnerability to loss, as a component of a poorly developed autonomy, may facilitate the ease with which the psychobiological process of withdrawal is precipitated. Such withdrawal from social interaction in turn reduces peer reinforcement and may further increase the social isolation. The resulting vicious cycle can become a major contributor to the precipitation of depressive illness.

In human beings these basic psychobiological processes are infinitely modified by abstract conceptual thought. Our ability to interact with the environment around us which we both adapt to and modify in our own interest is a function of our particularly complex and vigilant central nervous system. It frequently becomes the meaning of the environmental event, not the event per se, that determines the degree of psychobiological challenge that we perceive. Hence the development of conceptual tools with which to describe intrapsychic process becomes essential if we are to understand the genesis of affective change and disorder in human beings. That is the subject of our next chapter.

References

1. White, E. B. *Charlotte's web.* New York: Harper and Row, 1952.
2. Parkes, C. M. *Bereavement: Studies of grief in adult life.* New York: International Universities Press, 1972.
3. Lorenz, K. *Man meets dog.* Translated by Marjorie Kerr Wilson. Penguin Books, 1964.

4. Burton, R. (1621). *The anatomy of melancholy*. New York: Vintage Books, 1977.
5. Freud, S. (1917). Mourning and melancholia. In *Collected papers*. London: Hogarth Press, 1950, pp. 152–172.
6. Lindemann, E. Symptomatology and management of acute grief. *American Journal of Psychiatry*, 1944, *101*, 141–148.
7. Marris, P. *Widows and their families*. London: Routledge and Kegan Paul, 1958.
8. Clayton, P. J. Bereavement. In E. S. Paykel (Ed.), *Handbook of affective disorders*. New York: Guilford Press, 1982 pp. 403–415.
9. Maddison, D., and Walker, W. L. Factors affecting the outcome of conjugal bereavement. *British Journal of Psychiatry*, 1967, *113*, 1057–1067.
10. Maddison, D. C., and Viola, A. The health of widows in the year following bereavement. *Journal of Psychosomatic Research*, 1968, *12*, 297–306.
11. Harlow, H. F. The nature of love. *American Psychologist*, 1958, *13*, 673–685.
12. Harlow, H. F., and Harlow, M. K. Social deprivation in monkeys. *Scientific American*, 1962, *207*, 136.
13. Harlow, H. F., and Harlow, M. K. The affectional system. In A. M. Schrier, H. F. Harlow, and F. Stollnitz (Eds.), *Behavior of nonhuman primates*, vol. 2. New York: Academic Press, 1965.
14. Bowlby, J. *Attachment and loss*. I. *Attachment*. New York: Basic Books, 1965.
15. Bowlby, J. *Attachment and loss*. II. *Separation*. New York: Basic Books, 1973.
16. Bowlby, J. *Attachment and loss*. III. *Loss*. New York: Basic Books, 1980.
17. Lorenz, K. *Studies in Animal and Human Behaviour*, vol I. Translated by Robert Martin. Cambridge, Massachusetts: Harvard University Press, 1970.
18. Freedman, D. A., and Brown, S. L. On the role of coenesthetic stimulation in the development of psychic structure. *Psychoanalytic Quarterly*, 1968, *37*, 418–438.
19. Spitz, R. A. Anaclitic depression: An inquiry into the genesis of psychiatric conditions in early childhood. *Psychoanalytic Study of the Child*, 1946, *2*, 313–347.
20. Robertson, J., and Robertson, J. Young children in brief separation: A fresh look. *Psychoanalytic Study of the Child*, 1971, *26*, 264–315.
21. Holmes. T. H., and Rahe, R. H. The social readjustment rating scale. *Journal of Psychosomatic Research*, 1967, *11*, 219–225.
22. Paykel, E. S. Life events and early environment. In E. S. Paykel (Ed.), *Handbook of affective disorder*. New York: Guilford Press, 1982, pp. 146–161.
23. Schless, A. P., Schwartz, L., Goetz, C., and Mendels, J. How depressives view the significance of life events. *British Journal of Psychiatry*, 1974, *125*, 406–410.
24. Glassner, B., and Haldipur, C. V. Life events and early and late onset of bipolar disorder. *American Journal of Psychiatry*, 1983, *140*, 215–217.
25. Paykel, E. S., Myer, J. K., Dienelt, M. N., Klerman, G. L., Lindethal, J. J., and Pepper, M. P. Life events and depression: A controlled study. *Archives of General Psychiatry*, 1969, *21*, 753–760.
26. Brown, G. W., and Harris, T. O. *Social origins of depression*. London: Tavistock Press, 1978.
27. Lloyd, C. Life events and depressive disorder reviewed. II. Events as precipitating factors. *Archives of General Psychiatry*, 1980, *37*, 529–535.
28. Cadoret, R. J., Winokur, G., Dorzab, J., and Baker, M. Depressive disease: Life events and onset of illness. *Archives of General Psychiatry*, 1972, *26*, 133–136.
29. Ambelas, A. Psychologically stressful events in the precipitation of manic episodes. *British Journal of Psychiatry*, 1979, *135*, 15–21.
30. Kennedy, S., Thompson, R., Stancer, H. C., Roy, A., and Persad, E. Life events precipitating mania. *British Journal of Psychiatry*, 1983, *142*, 398–403.
31. Crook, T., and Eliot, J. Parental death during childhood and adult depression: A critical review of the literature. *Psychological Bulletin*, 1980, *87*, 252–259.
32. Engel, G. L., and Reichsman, F. Spontaneous and experimentally induced depressions in an infant with gastric fistula: A contribution to the problem of depression. *Journal of the American Psychoanalytic Association*, 1956, *4*, 428–452.

5

The Meaning of Loss: Psychoanalytic Explorations

Every organism, even the lowest, is not only in a vague sense adapted to, but entirely fitted into its environment. According to its anatomical structure it possesses a certain "Merknetz" and a certain "Wirknetz"—a receptor system and an effector system.

Yet in the human world we find a new characteristic which appears to be the distinctive mark of human life. The functional circle of man is not only quantitatively enlarged; it has also undergone a qualitative change. Man has as it were, discovered a new method of adapting himself to his environment. Between the receptor and the effector system we find in man a third link which we may describe as the symbolic system. This new acquisition transforms the whole of human life. . . . Man lives not merely in a broader reality, but in a new dimension of reality.

Ernst Cassirer[1]
An Essay on Man (1944)

The Roots of Psychoanalysis

During October, 1896, in Vienna, Austria, a retired wool merchant aged 81 years died following a short illness. Surviving him were a wife, 20 years his junior, and seven children. Shortly after his death, the eldest son, a physician, wrote to a friend, "My father's death has affected me profoundly—with his peculiar mixture of deep wisdom and fantastic lightness he had meant very much in my life. He had passed his time when he died, but inside me the occasion of his death has reawakened all my early feelings and now I feel quite uprooted."[2]*

Sigmund Freud, the writer of this letter, was born in 1856 in Moravia, now Czechoslovakia, the first child of Jacob Freud and Amalie Nathonsolm. Jacob, then age 40, had just married for a third time. He had once made a fine living in the wool trade. Increasing industrialization that spread from England in the early part of the 19th century, however, had dramatically changed the economy of the small country towns

---

* Letter to Fleiss.

of Europe. Such was the case in Freiburg, where the Freud family lived. Migration of individuals to the industrial centers had eroded the labor market and rapid inflation placed Jacob's livelihood increasingly in jeopardy. The final blow came in 1859. A new railway line was being built out from Vienna, and many merchants had pinned their hopes upon this transport for an economic revival of the region. The decision was made, however, to bypass the town of Freiburg. The following year, Jacob Freud moved his family to Vienna.

Times were initially hard, but industry and thrift produced an adequate living. Upon the eldest son, both parents placed great hopes. He was clearly an intelligent boy and it was determined that, to avoid those problems faced by his father, he should have a profession. Preferably he should become a physician. So, at some family sacrifice, Sigmund was given a room of his own, an oil lamp by which to study, and the best schooling available. Perhaps even more important, his mother offered a favored and warm closeness to her eldest son. Freud later felt that his great drive to success and a willingness to labor against opposition stemmed from this very special bonding that he had with his mother. "A man," he wrote, "who has been the indisputable favorite of his mother keeps for life the feeling of a conqueror."[2]

At the time of his father's death, Freud was already a controversial figure in Viennese medical circles. Earlier that year, on the second of May, he had presented a paper to the Society of Psychiatry and Neurology in Vienna entitled *The Etiology of Hysteria*.[3] It was an outline of his thesis that sexual factors play an essential part in the origin of the neuroses, particularly hysteria. Professor Krafft-Ebing, who chaired the meeting, declared afterwards that the paper was a "scientific fairy tale." Otherwise, Freud's presentation was met with silence.

Freud had started his specialist career a decade earlier as a neurologist with a traditional interest in neuropathology. He had some minor success with a gold chloride stain, a technique with which he had been able to obtain a "wonderfully clear and precise picture" of neuronal cells and fibers. At 29 he had gone to visit the Salpêtrière, the hospital in Paris from which Charcot dominated neurology in the late 19th century.

It had taken Freud little time to realize that Charcot's preeminence was entirely justified[4]; after two brief months, he persuaded Charcot to allow him to translate the great neurologist's most recent work into German. The laboratories in Paris were not as well equipped as those in Vienna, and Freud turned away from his interest in anatomic stains to clinical matters, becoming particularly interested in Charcot's teachings on hysteria. Upon returning to Vienna, Freud's desire to marry gave an additional reason for moving closer to clinical medicine; he felt that the laboratory and its preoccupations were incompatible with an adequate family life. In addition, to provide the necessary financial support for a family, Freud had to practice neurology as a clinical discipline.

Freud was in his early 30s when he settled in Vienna to begin his researches into the clinical aspects of hysteria. That he was still very interested in neuroanatomic structure, however, is reflected in his first book, published in 1891.[5] It is a discussion of the complex anatomy then thought to underlie the clinical phenomena of aphasia. Providing a detailed analysis of the neuropathology of published cases, Freud showed that the schemata that most authorities invoked to support their theories had inherent contradictions. He sharply questioned the then dominant theory that different clinical forms of aphasia could be explained by small subcortical lesions interfering with different neuronal pathways. In place of this localization, Freud introduced a functional explanation, a hypothesis more closely aligned with physiology and with Hughlings Jackson's vision of the nervous system than with that of contemporary anatomists.[6]

His principal interests, however, were increasingly focused upon hysteria and attempts to alleviate the condition through hypnosis. Freud's visit to Charcot had been stimulated by his conversations with Breuer, another Viennese physician. The focus for their common interest had been the case of Anna O., a person treated by Breuer for 2 years using hypnotic techniques. In general hysterical illness was considered a disorder of women and not worthy of medical attention in 19th century Viennese circles. Yet Charcot's systematic study had demonstrated the condition in both men *and* women. Most impressive of all, he had been able, by the use of hypnotism, to elicit hysterical symptoms, tremors, paralysis, anesthesias, and other symptoms that mimicked spontaneous hysteria in every detail.

Hence, on his return to Vienna in 1886, Freud was fired with considerable enthusiasm for the hypnotic technique. However, the practice of hypnosis he found unsatisfactory. He was not a good hypnotist himself and encountered resistance in his patients when using the technique to explore the origins of some of their symptomatology. Dissatisfied, Freud sought another solution. He began to experiment with a new method, that of free "association" to conscious events. Freud had been impressed by Charcot's thought that underlying many hysterical symptoms were sexually related anxieties. He became increasingly convinced of the truth of this when, using his new technology, the patient's blockade against recall fell away, frequently revealing that the focus of concern was sexual thoughts and happenings. To describe the subsequent evolution of Freud's work would demand greater detail than is appropriate in our discussion of mood disorders. However, a decade later, at the time of his father's death in 1896, Freud had both developed the framework of his new technique and also the concept of unconscious mental life.

Jacob Freud's passing became the stimulus for considerable introspection on the part of his eldest son. After the death Freud embarked upon "self-analysis," including the interpretation of his own dreams. There is some evidence that the bereavement had left him feeling miserable. He described himself as "in a bad mood," and the self-examination he undertook in part to mitigate the melancholy influence upon him. This intensive period consolidated many of Freud's ideas. He reflected upon the meaning of his fascination with his mother and discovered in his own dreaming what he believed to be incestuous wishes toward his children. In his writing he equated the symptoms of anxiety with the physiological changes of sexual intercourse. That accelerated breathing, palpitations, and sweating appeared in both reinforced his thoughts that sexual motivation underlay the symptoms of hysteria. A more profound insight, however, came early in 1897, when he realized that many of the sexual preoccupations of his patients were actually fantasy and did not equate with experience. This led him into an exploration of the sexual life of the child and what he believed to be the erotic nature of the oral and anal areas during the first few years of life.

These historical events provide an important background for those seeking to understand the evolution of theories of intrapsychic function in mood disorders. Sigmund Freud did not actually invent many of the ideas for which he has become famous, for concepts of unconscious mind and the technique of free association to bring forth hidden information can be traced to earlier writers. However, his writings and the psychoanalytic movement that he began have had a profound effect upon the language we use to describe the dynamics of intrapsychic phenomena. What began as an interest in psychopathology has emerged as a broadly accepted theory of symbolic mind itself.

Early Psychoanalytic Concepts of Melancholia

Freud had almost certainly learned in high school of Herbart's* doctrine of the unconscious. However, through his experiences as a clinician and from what he had gathered through his association with Charcot, his picture of the unconscious differed considerably from that of his predecessors. Rather than a storage house or a repository of old ideas and memories, he considered it a dynamic force, one which played a major role in mental life. He saw the energy of the unconscious coming from the instinctual drives ("triebe"), particularly the sexual drive. Unconscious strivings were kept largely out of awareness because they were a threat to the individual's smooth social functioning. They were, in fact, actively repressed. Thus Freud's theories sought to describe the homeostasis of psychological mechanisms. As socially unacceptable drives disturbed the psychic equilibrium, detour behaviors occurred which inhibited the drive to serve the greater well-being of the total individual. Neurotic symptoms were an unstable compromise between attempts at drive gratification and learned components of the individual's personality—particularly the incorporated parental image which Freud later termed the "superego."

It was some time before the early theorists turned their attentions to disorders of affect. Karl Abraham was by then a dominant figure in Freud's circle of friends and associates. He is described by Ernest Jones, Freud's biographer, as the most normal member of the group. "His distinguishing attributes were steadfastness, commonsense, shrewdness, and perfect self-control. However stormy or difficult the situation, he always appeared to retain his unshakable calm. Abraham would never undertake anything rash or uncertain—he was the most reserved, the least expansive of us."

It was Karl Abraham who first undertook psychodynamic elaboration of the melancholic state. Like all his disciples Abraham was greatly influenced by Freud's views on infantile sexuality, which inferred that all pleasure and eating tied directly to a pregenital stage of sexual life. Hence, many of Abraham's writings explore the appetite disturbances of manic–depressive individuals. Both the melancholic's inability to eat and the tendency to overeat seen in individuals with cyclic depression were explained symbolically. The melancholic directed toward the object of his sexual desire the wish to incorporate, to devour, and to demolish. Abstention from food punished the depressive for "unconscious cannibalistic drives." In other depressed persons the sustenance provided to the infant by the mother might serve later in life as a

---

* Johann Herbart, the influential German philosopher and educator who was born in 1776 and held the Kant Chair of philosophy at Konigsberg between 1809 and 1833. He died in 1841, but his philosophy was taught widely in German high schools during the late 1800s.

comforting substitute for mothering itself. In this way Abraham sought to explain the overeating characteristic of some bipolar depressed individuals.[7]

A more general analysis of the melancholic state was provided by Freud himself in his classical paper *Mourning and Melancholia,*[8] published in 1917. Other papers were developed by Abraham, both before and after Freud's own interpretation.

Both writers attempted to define the difference between psychotic depression, or melancholia, and neurotic illness and mourning. Freud described the following elements as characteristic of melancholia:

- A profoundly painful dejection
- Abrogation of interests in the outside world
- The loss of the capacity to love
- The inhibition of all activity
- The lowering of self-regard to the extent that self-reproach and a delusional expectation of punishment occur

Freud then drew three general distinctions between mourning and melancholia:

1. Unlike mourning, melancholia occurred only in specially predisposed individuals.
2. The melancholic may not, in fact, have lost the object of his love, but merely perceived it as lost.
3. In mourning, self-respect (or ego) remains intact, whereas in melancholia, the individual turns in upon the self with blind hatred.

In 1911, Abraham had postulated that the melancholic's destructive self-image was an extension of the belief that he was unworthy and defective, unlovable and incapable of loving. The depressed individual oscillated between despair at not being loved by others and hatred of them, because it was perceived that they did not return the affection.

Freud extended this concept of ambivalence, suggesting that the hatred towards the "loved object" was unacceptable to the melancholic because of its destructive force. Thus, hatred was "introjected" or turned as a reproach against the self. This was the mechanism underlying those psychotic elements of the illness that led to self-destructive thoughts and acts. Furthermore, Freud argued, individuals prone to depression tend to choose lovers whom they perceive as being very similar to themselves, fostering the ambivalence.

Both Freud and Abraham cautioned against equating psychodynamic explanation in melancholia with the cause of the disorder itself. Cognizant of the many somatic components of the illness, they believed that genetic and neurobiological research would be necessary to explain the particular predisposition that some individuals have to melancholia. Subsequently such cautionary notes appear to have been lost as the exploration of intrapsychic life became a goal unto itself.

Structural Analytic Approaches

Freud's *Mourning and Melancholia* is the seminal analytic paper from which we now move in our discussion. However, it was written before Freud had outlined his structural theories of psychic life. The first comprehensive exposition of these theories appeared in his book *The Ego and the Id,* published in 1923.[9]

In 1928 Sandor Rado restated the early psychoanalytic concepts of depression in the new language of id, ego, and superego.[10] Rado viewed the depressive as possessed of an "intensely strong craving for narcissistic gratification and tremendous narcissistic tolerance." The depressive is viewed as dependent for the maintenance of ego function upon external sources of love and approval. Those predisposed to melancholia constantly seek evidence that others perceive them as lovable. However, once this adoration is received, the loving person is treated as a possession. Then when, intolerant of such captivity, the love "object" withdraws, the depression-prone person becomes angry and bitter. Desperate attempts are made to regain the lost love. Should these efforts fail, there swiftly follows superego punishment and remorse. Melancholia, described by Rado as "a great despairing cry for love," then ensues. When the ego moves from reality to a level of fantasy or the development of "psychic institutions" in an attempt to regain the loved one, then by definition, psychosis has occurred.[11]

Rado's writings are difficult to follow, but they mark a transitional step in psychoanalytic writing from the pure libidinal theories of Freud and Abraham to structural concerns and the central issue of self-esteem. Moreover, the formulation does have clinical reference, for exaggerated dependency and the desperate seeking of approval from others are well-known clinical phenomena in depression-prone individuals. In many such cases the individual develops attachments largely on the basis of what the other person can do to define and bolster the depressive's self-esteem. Metaphorically, the activity has been described as shopping for "narcissistic supplies." While to some extent all persons indulge in such activity, Rado saw its dominance as evidence of pathology. When relationships (attachments) are based primarily on the ability of one partner to bolster the other's self-esteem, there is inherent instability. Considerable "acting out" behavior often ensues, unconsciously motivated by the desire to be rejected and to confirm the distorted self-image. Such behavior, at the same time, expresses an unconscious anger stemming from an awareness of unreasonable dependence on others.

Thus Rado broadened Abraham's and Freud's earlier themes of ambivalence and introjection to include structural concepts of narcissistic and dependent craving for affection by the ego. A punitive superego is seen as responsible for the self-accusatory behaviors of the melancholic.

The Mother–Child Relationship and the Depressive Position

The early analytic literature frequently equates depression with an obsessive character structure. These ideas broaden the discussion of oral and anal cathexis begun by Freud and Abraham. The underlying thesis is that, during infancy, neurotic preoccupations progress through the oral stage, but become fixated at the anal stage of development. Later in life this is reflected as a preoccupation with order and cleanliness.

Gero, who did not subscribe to Freud's ideas of psychosexual development, did, however, seek to broaden the concept of orality in depressive illness. In two patients he undertook a comprehensive analysis of the mother–child relationship and concluded that "oral pleasure is only one factor in the experience satisfying the infant's need for warmth, touch, love, and care."[12] We are reminded of Harlow's pioneering primate studies and the evidence reviewed earlier that infant–mother attachment systems probably reflect a primary drive.

While this part of Gero's theory foreshadows many later developments, other elements are more abstract than is clinically useful. He was perplexed by the depressive's persistent infantile demand for love and the seeming inability of such individuals to acquire more mature ways of receiving and giving affection. Anxiety around genital sexuality which repeatedly returns the libidinous energy to a pregenital stage is offered as a probable explanation. These concepts apparently influenced the ideas of Melanie Klein.

Klein focused her analytic interest in depression on the infant's first year of development. Her formulations are always complex and at times mystical, with little attention to neurobiological development. She theorizes that infants feel a strong aggressive drive during the first year as a reaction to the frustration and lack of gratification that they all experience. Recognizing this inner rage, the infant fears extermination and feels helpless as a result. Projection of these aggressive feelings to the most immediate external object, the mother's breast, results in the infant's identification of the breast (an extension of the maternal image) as a source of both gratification and persecution. Klein terms this stage of development the "paranoid position."[13]

In the course of time, other parts and behaviors of the same mother—"good objects"—are also introjected by the infant as needs are satisfied. A warm emotional tie develops. However, this process takes time, and may not be satisfactorily resolved. Then oscillation between paranoid rage and frustration and the alternate sense of gratification and pleasure leads later to pathological behavior. Klein believed that children who are insecure in being loved do not internalize the good "objects" and are predisposed to the "depressive position" in later life. Such individuals struggle with feelings of loss, sorrow, guilt, and a lack of self-esteem.

Discarding the symbolic references contained in Klein's ideas and recasting them in the language of bonding theory and attachment systems, we again find considerable resonance with recent infant research. Klein was one of the first psychoanalytic writers to emphasize that predisposition to depression depended not merely upon traumatic incidents but on the overall quality of the early mother–child relationship. If secure bonding develops, then the infant feels loved and is more able to adapt to absences of the mother. A sense of self-esteem and an ability to survive alone is fostered through the bonding experience. When this basic attachment system is distorted, the infant or child may come to feel "paranoid," i.e., undervalued, insecure, and depressed. Later separations or losses then have powerful implications, threatening the maintenance of self-esteem.

The Evolution of Analytic Concepts

If we pause now to evaluate the progress of our review, we may reflect that the current "models" of intrapsychic depression which were outlined in Chapter 2 are drawn from general psychoanalytic concepts. The models incorporate elements from what was in reality a continuous evolution of ideas. The early theorists saw the mind as a closed system; more recently the individual or ego is conceptualized as in balanced commerce with external forces. With time there has been a broadening of the language used; most theorists, while adhering generally to Freud's structural theories, have developed words and phrases that reflect slightly differing views of similar phenomena. Inevitably, theoretical debates have arisen. For example, is the maintenance of self-esteem an ego function or is it dependent upon the balance of forces between the theoretical structures of ego, id, and superego? In the next section, we refer briefly to these discussions while trying to avoid the semantic arguments. We confine our review to issues of self-esteem, the ego, and depression.

Self-Esteem, the Ego, and Depression

In an extensive review of the evolution of structural psychoanalytic theory, Meyer Mendelson emphasizes the pitfall of accepting the idea of structure too literally. The division of the self into component parts is, after all, an abstract tool. Logically and coherently, it is essential to focus on the *person* as depressed and not the ego or any component of the abstract structure. The terms ego, superego, id, self-esteem, and ego ideal "are ordering abstractions, theoretical devices to classify data (and ideas) according to semi-arbitrary, but clinically useful categories. They are not mutually exclusive structures, nor do they contain affects, defenses, or self-representations."[14]

Freud, in 1917, in *Mourning and Melancholia,* wrote that the self-accusatory aspects of melancholia could be viewed as "the conflict between one part of the ego and its self-criticizing faculty." This "faculty," a controlling conscience, was to evolve into the concept of the superego. Earlier, when discussing narcissism, Freud had distinguished the notion of ego ideal from that of conscience, suggesting that the former grew from a set of standards that an individual set up to measure personal achievement. Superego, on the other hand, was defined as that psychic structure that enforced the standards of conscience upon ego activity.

In the early 1950s a series of proposals, now generally accepted, more clearly defined the superego as the conscience element within the self and the ego ideal as those self-images, conscious or unconscious, to which any individual aspires. Self-esteem can then be best understood as representing the degree of discrepancy perceived by the individual between the preferred concept of self and that self-representation accepted as the realistic one.

Edward Bibring,[15] writing during the 1940s and early 1950s, and Edith Jacobson,[16] whose contributions span a quarter of a century from the mid-1940s, have both been interested in the theoretical underpinnings of depressed mood and the relationship of depressive phenomena to this concept of self-esteem.

Bibring believed that the mood of the depressive reaction could best be understood as a result of the lowering of self-esteem. He attempted to broaden the earlier psychoanalytic view that depressive reactions were a direct result of traumatic childhood experience. Rather, the traumatic event disrupted the usual working-through of oral needs, resulting in fixation of psychological development. When subsequent events threatened oral (narcissistic) supplies, the infantile state of helplessness and powerlessness returned. Not just the withdrawal of oral supplies but the frustration of any narcissistic aspiration could precipitate a loss of self-esteem and the associated depressive mood. Thus symbolically, frustration of "oral" aspirations (a wish to be good, kind, or clean) might be compounded by the frustration of "phallic" strivings, such as the wish to be strong, superior, great, and secure. In his theoretical structure, Bibring placed great emphasis upon depression as a disorder of the ego itself, rather than conflict between the ego and the superego. As Mendelson points out, this semantic detail, while perhaps important at the time, creates a theoretical red herring dependent upon whether ego ideal is conceptualized as part of the ego or part of the superego. The construct thus varies with the theorist.

Edith Jacobson's theories offer a complex dynamic approach to normal self-regard, mood disorder, and cyclothymia. Her work is of particular interest because its developmental focus offers a bridge to Bowlby's observations of human infants and to studies of primate behavior.

Like Bibring, Jacobson emphasizes the loss of self-esteem as central to understanding the intrapsychic mechanisms of clinical depression. The development and maintenance of self-esteem are tied to the fundamental goal of integrating the self-image, including the distinction of the self as an autonomous individual. Self-image is the awareness of the self as a biological, functional, thinking organism. It is distinguished from "self-representation," which is the specific intrapsychic view of the self, and "object-representation," which is reserved for describing individuals and objects outside the person. Successfully achieving integration of the self-image occupies early childhood development and continues well into adult life. Initially, of course, there is no specific awareness of the self, and confusion exists between the sensations that are received from the infant's own body and those of the immediate environment. (Here we see a similarity to some of Klein's views.) The distinction between the two is a task optimally accomplished in the atmosphere of a supportive and nurturing environment. A stable bond between parent and child promotes successful adaptation to perturbing stimuli, without the infant being prematurely overwhelmed. Inconsistent or inadequate mothering places the infant in this danger and frustrates the adaptive striving, while overindulgence leads to a slowing of boundary definition between the self and the external world.

Jacobson adheres to the classic psychoanalytic view that it is the cathexis, the psychic energy invested, in the self and the external world that ultimately determines self-esteem. This cathexis may be libidinal or aggressive (of love or hate), but usually is a mixture of the two, thus providing the infantile roots of ambivalence. As the child grows selective imitation of the parents and other important persons in the environment takes place. This heralds the development of an identity based on personal values (self-regard). Progressive distinction from the parent occurs as these ego functions of discrimination and judgment mature. The ego ideal develops both as an idealized object and as a preferred self-image. Self-esteem, defined as the degree of discrepancy between the self-representation (the intrapsychic view of the self) and the preferred concept of the self (the ego ideal), then becomes an entity for care and maintenance. The greater the positive cathexis of the self and the better the distinction between the self-representation and the object-representation, the greater is the individual's resilience and ability to overcome changes that threaten self-esteem in later life.

Jacobson argues that anything that interferes during the child's development with this progressive distinction and identification with love objects leads to a very unstable and poorly integrated superego. This latter may then assume excessive control over the day-to-day functions of the ego and the individual's behavior. Superego dominance, reflected as a need for order and control, has been described as characteristic of many predepressive personalities. Loss of control, withdrawal of love, the falling short of a goal that is predetermined by the ego ideal then all become capable of generating a rapid and profound lowering of self-esteem, the consequence of which is depressive illness.

Theories of Manic–Depressive Oscillation

As in other areas of our review, we find that mania has received comparatively little attention. A few theoretical papers have appeared, notably by Deutsch[17] and Lewin.[18] Both writings were based upon the analysis of patients who had passed through a hypomanic episode. One study frequently cited is that of Cohen and colleagues, who studied the family backgrounds of a group of manic–depressive individuals in analysis. They concluded that the families perceived themselves as different from their neighbors, striving constantly to enhance their own family prestige. Usually one child is chosen as the family standard bearer, and parents' approval of the child's behavior becomes implicitly dependent upon the child's accomplishments. Cohen and associates believed that it was this child who later developed bipolar illness. Fathers of the families were described as unsuccessful and weak, while the mothers were ambitious, aggressive, strong, and stable. The patients appeared to be most closely identified with their father. The researchers suggested that this variation from the customary parental model provided an unstable identification for the child. A poor ego ideal and an unsatisfactorily integrated superego mechanism was the result.[19]

Other evidence to support Cohen's study is scant. Most observers are doubtful that any specific pattern of family interaction is present during childhood for those individuals who develop bipolar illness. Other psychoanalysts do maintain that the principal psychodynamic mechanism in mania is denial of a very brittle or poor self-image, however. Manic episodes are seen as a merging of the ego functions of the individual with the superego, so that transiently the energy of the latter provides a righteous self-inflation of the ego, with euphoria and unbridled narcissistic behavior. As the denial of the real view of the self breaks down, depression results. These theoretical mechanisms appear to have little practical value to the clinician, save perhaps for identifying denial as a major psychic defense in mania. Common experience does suggest that manic illness is frequently precipitated by what the individual perceives retrospectively as a loss or undesirable episode.

Reality Distortion in Depression

Most analytic writers implicitly acknowledge the distinction between neurotic and psychotic depression. Conceptually, psychosis is viewed as ego regression to an earlier stage of psychological development. The psychotic individual is seen as replacing the normal adult object relationships with earlier ones which exist entirely within repressed experience. In the symbolic language of Rado,[11] "the weakly ego gives up its struggle with the world and sets up its own psychic institutions." These psychic institutions usually distort what other persons believe to be the individual's attributes. The preoccupations of the lawyer cited in our first chapter would be an illustration of such distortions, which are generally accepted but not usually interpreted as secondary to psychological regression.

Aaron Beck has focused on such cognitive distortions as the principal psychological disturbance in depression and has formulated a theory of cognitive treatment based upon them.[20–22] While such distortions are most floridly seen in psychotically depressed people, Beck maintains that the depressive sees himself, his world, and his future in an idiosyncratic way even when not in a psychotic state. The triad of a negative interpretation of experience, of the self, and of the future forms a depressive core which is the focus for a psychotherapeutic intervention named cognitive therapy. Because of this cognitive disability, Beck argues that mood disorders cannot be considered purely a disturbance of affect. His treatment paradigm does have the major advantage of being quantifiable and provides the basis for specific therapeutic technique. Early results using this therapeutic rationale have been encouraging in nonmelancholic depressives.

Some Conclusions and Comment

The early career of the driven, ambitious Freud, eager to develop a new clinical technique, stands in contrast to the abstract constructions of mental life that have marked psychoanalysis in later years. What began as an effort to explain clinical phenomena appears, at times, to have become endangered by doctrinaire pronouncements. Some indeed dismiss psychoanalytic investigation, with its constant reference to the previous literature, as more akin to the justification of religious dogma than to science. Unfortunately, the discipline of psychoanalysis has been very tolerant of theory without adequate objective data to support it. The defensive argument frequently put forward is that the data can only be gathered in a very personal and laborious way. The counter-argument is that such statements cannot serve as an apology for a lack of investigational rigor.

It is possible beyond doubt for all disciplines to become lost in their own linguistic forms, and care must be exercised to confine claims of clinical validity in psychoanalysis to those warranted by observation and outcome. The theory has provided, however, a valuable structure and dynamic language for exploring the complexities of intrapsychic life. That alone has been an enormous contribution.

It was Freud's original intention to develop a psychology for neurologists. He was heavily influenced by the writings of Helmholtz, the eminent German neurophysiologist, physicist, and inventor of the ophthalmoscope, who died in 1894. Helmholtz eschewed the theory of a supraorganic vital force as an explanation of nervous system function and sought to replace it with a mechanical one. Probably Freud's theories of ego defense and adaptation which evolved over decades stemmed from these early concepts of self-regulation and cybernetics. Initially, however, Freud failed to see the interaction of the individual and the environment as a closed loop (see Chapter 8) and thus developed the theory of drive or instinct (triebe) to explain the motivation behind his patients' actions. His search was for a dynamic construct, more in keeping with physiology than neuroanatomy.

Psychodynamic theories of mood have always grown as part of the concern with general psychoanalytic concepts. In that the interest has been largely the symbolic explanation of mood states rather than the major depressive syndrome, this is appropriate. The dynamic construct of self-esteem or self-worth is particularly useful. This notion allows one to build on the behavioral observations in human infants and primates and develop a viable theory of human psychic life. The early bonding which we discussed in the last chapter is infinitely modified by the symbolic representation, the unique templates we build of ourselves and our world. These templates, generalizations about what we have experienced and upon which our future behavior rests, are not always accurate. While they are modified by experience throughout life, the quality of our success in early interaction with important figures plays a disproportionate role.

Should early aspiration, our ego ideal, be distorted and poorly integrated with true ability, then self-esteem (or self-worth), as a measure of the difference between that to which we aspire and that which we can achieve, will remain unstable. Adaptive styles develop accordingly: unquestioning dependence upon those dominant around us, excessive control of our environment, these rescue operations become a substitute for a valued self. The usual exigencies of life—such as loss of affection, complex demands,—then become events of major psychobiological challenge. The capacity to adapt smoothly to the demand is compromised, and we exist in a chronic state of heightened arousal.

Exploration of our physical and intellectual abilities within a nurturing, controlled environment where adaptive demand is graduated appropriately to psychobiological development is an essential early task. The antecedent correlates of self-esteem in a study by Coopersmith of 10–12-year-olds, have been shown to be parental warmth, acceptance, respect, and clearly defined limit setting. Self-judgment is more influenced by those in the immediate effective interpersonal environment than by public standards. Thus self-esteem and its regulation appear to be disproportionately related to the early experience of infantile attachment and parental reinforcement.[23] Epstein suggests that "a person with high esteem carries in effect a loving parent within him."[24] The reader will recall that Freud said something similar, and one suspects that it was autobiographical[2]: "A man who has been the indisputable favorite of his mother keeps for life the feeling of a conqueror."

References

1. Cassirer, E. *An essay on man: An introduction to a philosophy of human culture.* New York: Bantam Books, 1944.
2. Jones, E. *The life and work of Sigmund Freud,* abridged version. New York: Basic Books, 1961.
3. Freud, S. (1896). The etiology of hysteria. In *Collected Papers,* vol. I. New York: Basic Books, 1959, pp. 183–219.
4. Freud, S. (1893). Charcot. In *Collected papers,* vol. I. New York: Basic Books, 1959, pp. 9–23.

5. Freud, S. (1891). *On aphasia*. Translated by E. Stengel. London: Imago Publishing, 1953.
6. Bay, E. Sigmund Freud's contribution to the early history of aphasiology. In F. C. Rose, and W. F. Bynum (Eds.), *Historical Aspects of the Neurosciences*. New York: Raven Press, 1982.
7. Abraham, K. (1911). Notes on the psychoanalytic investigation and treatment of manic–depressive insanity and allied conditions. In *Selected papers on psychoanalysis*. London: Hogarth Press and The Institute of Psychoanalysis, 1927, Pp. 5–9.
8. Freud, S. (1917). Mourning and melancholia. *The complete psychological works of Sigmund Freud, standard edition*, vol. 14, London: Hogarth Press and The Institute of Psychoanalysis, 1975, pp. 243–258.
9. Freud, S. (1923). The ego and the id. *The complete psychological works of Sigmund Freud, standard edition*, vol. 19, London: Hogarth Press and The Institute of Psychoanalysis, 1961, pp. 1–59.
10. Rado, S. The problem of melancholia. *International Journal of Psychoanalysis*, 1928, *9*, 420–438.
11. Rado, S. Psychodynamics of depression from the etiologic point of view. *Psychosomatic Medicine*, 1951, *13*, 51–55.
12. Gero, G. The construction of depression. *International Journal of Psychoanalysis*, 1936, *17*, 423–461.
13. Klein, M. *Contributions to psychoanalysis, 1921–1945*. London: Hogarth Press and The Institute of Psychoanalysis, 1948.
14. Mendelson, M. *Psychoanalytic concepts of depression*. Flushing, New York: Spectrum Publications, 1974.
15. Bibring, E. The mechanism of depression. In P. Greenacre (Ed.), *Affective disorders*. New York: International Universities Press, 1953.
16. Jacobson, E. *Depression*. New York: International Universities Press, 1971.
17. Deutsch, H. Absence of grief. *Psychoanalytic Quarterly*, 1937, *6*, 12–22.
18. Lewin, B. D. *The psychoanalysis of elation*. New York: W. W. Norton, 1950.
19. Cohen, M. B., Baker, G., Cohen, R. A., Fromm–Reichman, F., and Weigert, E. W. An intensive study of twelve cases of manic–depressive psychosis. *Psychiatry*, 1954, *17*, 103–138.
20. Beck, A. T. *Depression: Clinical, experimental, and theoretical aspects*. New York: Paul Hoeber, 1969.
21. Beck, A. T. The development of depression: A cognitive model. In R. Friedman and M. Katz (Eds.), *Psychology of depression: Contemporary theory and research*. Washington, D. C.: Winston-Wiley, 1974.
22. Beck, A. T. *Cognitive therapy and emotional disorders*. New York: International Universities Press, 1976.
23. Coopersmith, S. *The antecedents of self esteem*. San Francisco: Freeman, 1967.
24. Epstein, S. Anxiety, arousal and self concept. In L. G. Sarason and C. D. Speilberger (Eds.), *Stress and Anxiety*, vol. III. Washington, D. C.: Hemisphere, 1976.

The Neurobiological Foundations of Behavior: Environmental Challenge and Response

Admitting that vital phenomena rest upon physicochemical activities, which is the truth, the essence of the problem is not thereby cleared up, for it is no chance encounter of physicochemical phenomena which constructs each being according to a pre-existing plan and produces the admirable subordination and the harmonious concert of organic activity. There is an arrangement in the living being, a kind of regulated activity, which must never be neglected because it is in truth the most striking characteristic of living beings. . . .

Vital phenomena possess indeed their rigorously determined physicochemical conditions; but at the same time they subordinate themselves and succeed one another in a pattern and according to law which pre-exists. They repeat themselves with order, regularity, constancy, and they harmonize in such a manner as to bring about the organization and growth of the individual animal or plant.

Claude Bernard[1]
Introduction to the Study of Experimental Medicine (1865)

Introduction

This chapter is particularly addressed to those readers who do not have a background in medical or biological science. It offers first a brief selective survey of neurobiological and biobehavioral research which we hope will provide a useful technical background to our discussion in Chapter 7 of the neurobiology of mood. Subsequently we introduce the reader to the literature on the psychophysiology of environmental challenge.

There is something peculiarly attractive about information gained from biological investigation; even in the study of behavior, for many, it implicitly states something special about cause. There is no really satisfactory explanation for this. Perhaps knowledge derived through a complex technology, which is increasingly difficult for the

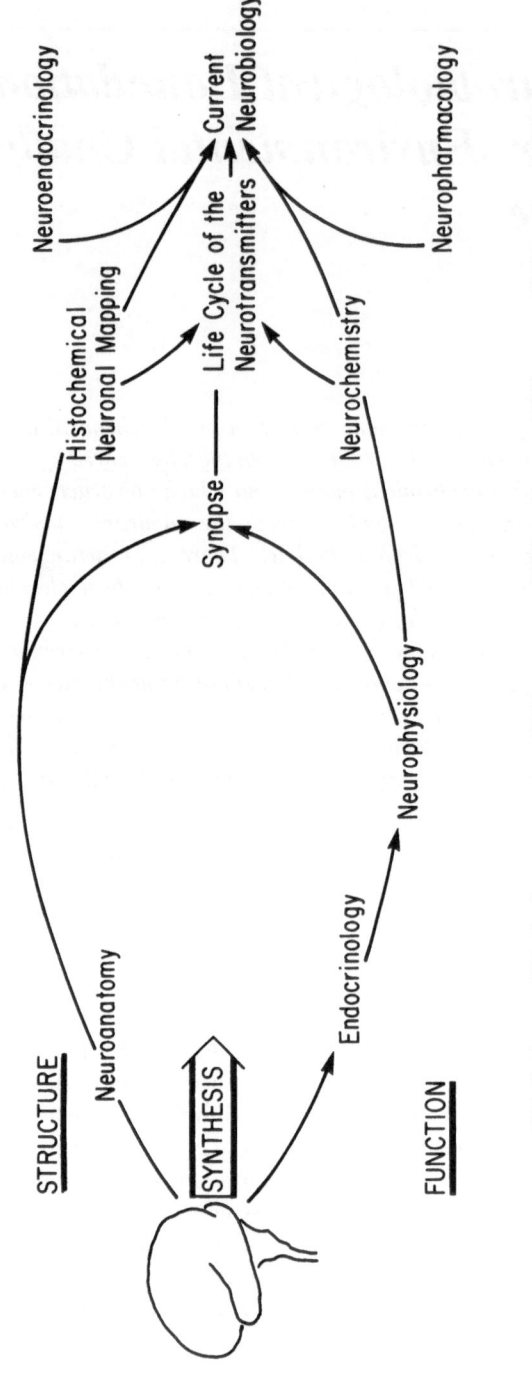

Figure 6.1 Historical roots of modern neurobiology: traditional divisions of structure and function are breaking down.

generalist to understand, has greater credibility than that gained through simple observation, a skill that everybody believes to be within their ability.

It is important to remind ourselves that neurobiology is but one of a number of concomitant conceptual levels at which we seek information to better understand behavior. Over time we have adopted these divisions to aid in analysis of the whole. Within neurobiology the disciplines of neurochemistry, anatomy, physiology, and endocrinology have a historical integrity, with varied tools of inquiry and different bodies of knowledge as a result (Fig. 6.1).

Recently, however, with a greater understanding of molecular biology, concepts of structure and function have merged again; within the complex physical and homeostatic properties of the cell, we find that structure determines function and function in turn molds structure. Living mechanisms are a study of order in space and time.

Claude Bernard was probably the first physiologist to recognize that the secret to life is control of the immediate environment. Without stability in the intra- and extracellular fluids of a multicellular organism, the processing of chemical information in and out of the cell is impaired, and soon the whole creature becomes disordered and dies. All mechanisms in biology have this inherent property of maintaining order and stability through constant change. Within the animal the maintenance of this order, the *milieu organique intérieur*, confers an autonomy from the larger external environment. Then, if the behavior of the animal as a whole appropriately modifies or finds compromise with its larger world, cellular processes are preserved. The activities are interdependent. The technology necessary to understand biological processes indeed may be more complex, but the information derived should not be given primacy automatically over psychological, social, or ethological observations. Ideally, each feeds our understanding of the other dimensions.

Illustrations from Endocrinology

Chemical Signaling Systems

The endocrine system and the central nervous system share much in common as highly developed regulatory mechanisms for the orderly transfer of information between cells. Such regulation is key to the survival of all multicellular organisms.

Endocrine signaling systems appear early in evolution. A chemical substance or messenger is released from one cell, and when it arrives at the surface of a second, it modifies the latter's behavior. The central nervous system may be viewed simply as a highly adapted group of cells in which the same principle holds. The specialized shape of the neuron has facilitated close physical contact and more rapid transmission of the necessary information. The communication still depends upon the release of a chemical substance, however, which now passes across a very small gap (the synapse) rather than via the bloodstream, to modify the cell membrane of the second neuron. Many of the messengers in the nervous system (neurotransmitters) also appear as chemical messengers in the endocrine system, where they are termed hormones. In higher animals the special interrelationship between these two systems is particularly evident in the hypothalamus of the brain, where the autonomic and endocrine systems are integrated and the peptides that release the various endocrine hormones are manufactured.

Such systems of chemical endocommunication and control are widespread throughout the animal and plant world. The hormones that occur in man and in primates appear early in the phylogeny and in many instances are also present in plants. The nuclei of steroids, important hormones in humans, for example, when combined with carbohydrates in plants form glycosides. Digitalis, found in foxgloves and used in the treatment of heart disease, is an illustration of such a combination. More recently, small peptides (endorphins) occurring naturally in the nervous system have been found to be analogues of the opiate drugs.[2,3] During evolution, serial adaptation has exploited these messenger molecules to good advantage. The steroids, which play a major part in the maintenance of the body's electrolyte balance, discharge the same responsibility in fish. The exchange of ions and water between the gills of fish is under corticosteroid control. The transmission of sodium through the highly permeable skin of the frog and the production of "tears" (avian dew drops) by marine birds, who can live for many days without fresh water, is similarly controlled. We retain vestiges of such functions in our own salty tears and in the capability of the sweat glands to move sodium out of the body during perspiration.

Théophile de Bordeu, a fashionable physician at the court of Louis XV, was the first to suggest the theory of internal messengers. His objective was to confirm the Hippocratic theory of humoral pathology. He believed that the three Hippocratic stages of disease (irritation, coction, and crisis) were dependent upon glandular secretions. Noting that eunuchs behaved differently from noncastrated men, he proposed that the changes were secondary to a loss of internal secretion from the sex organs. Berthold, in 1849, confirmed his observations, showing that capon characteristics which otherwise would have followed castration were prevented when the fresh testes were transplanted to another part of the cockerel's body.

Homeostatic Control Mechanisms

It was Bernard, however, who appreciated that *control* mechanisms were necessary if the amount of messenger released was to be appropriate to need. His research into the mechanisms surrounding the storage and release of glycogen in the body is a monument in the conceptual evolution of endocrinology. It stimulated the formation of his famous principle of the *milieu intérieur* and led to early research on feedback control mechanisms of neuroendocrine function.

Neuroendocrine regulatory mechanisms have certain principles in common with all other biological control systems. The most important is the closed homeostatic loop in which the controlled variable itself is the factor determining changes in the rate of its manufacture. The feedback system most commonly employed is one in which the output of the system inhibits production—a negative feedback system. The brain–thyroid axis illustrated in Fig. 6.2 will be used as an example.

The production of thyroid hormone inhibits the secretion of pituitary thyrotropin (TSH) by negative feedback mechanism such as that just described. Hence, a rising

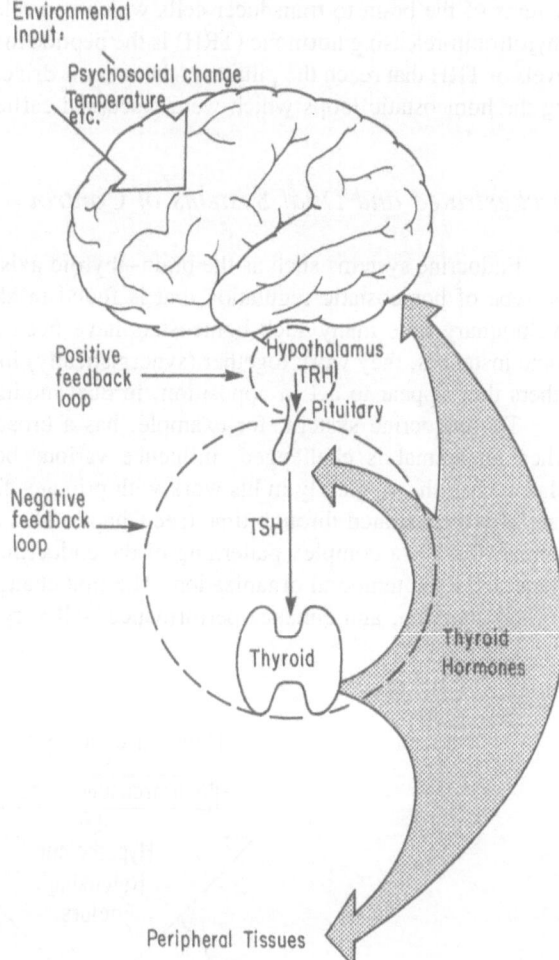

Figure 6.2 Control of a typical endocrine system: The brain–thyroid axis.

production of thyroid hormone inhibits TSH secretion and reduces the stimulus to further thyroid hormone production and release. However, it is probable that a positive feedback loop also operates when a rising output accelerates production. Evidence has been presented that such a system exists in the brain–thyroid axis, and similar mechanisms are found in the regulation of many cell enzymes. The time frame of response for the two mechanisms is different, however, with the positive loop having a shorter period than the negative one. This positive feedback then serves to maximize acute production which is ultimately inhibited through negative feedback to both the pituitary and the hypothalamus as peripheral levels of the hormone rise.[4]

The level of hormone production is modified by psychosocial and physical events through a set of mechanisms which ultimately integrate within the hypothalamus. Information from the environment is conveyed through the sensory and neurochemical systems of the brain to transducer cells which can release small peptide messengers. Thyrotropin-releasing hormone (TRH) is the peptide in our example. Depending on the levels of TRH that reach the pituitary, the latter is driven to release TSH, thus influencing the homeostatic loops which were discussed earlier.

Orchestrated and Dual Systems of Control

Endocrine systems such as the brain–thyroid axis provide a beautiful example of the type of homeostatic regulation that is found in all biological mechanisms. Over evolutionary time many such homeostats have been layered one upon the other. In some instances, they work together (synergistically) in temporal sequence, whereas in others they appear to act in opposition, in dual modification of similar mechanisms.

The endocrine system, for example, has a broad variety of subsystems which, when an animal is challenged, influence various bodily processes (cf. Fig. 6.3). Mason[5] has shown clearly in his work with primates that this response to challenge is carefully orchestrated through time (see Chapter 9). Following the perturbation of the animal, there is a complex patterning of the endocrine system which has a functional hierarchy and a temporal organization. The first changes are those that mobilize energy, aid attention, and enhance performance—all very necessary qualities under chal-

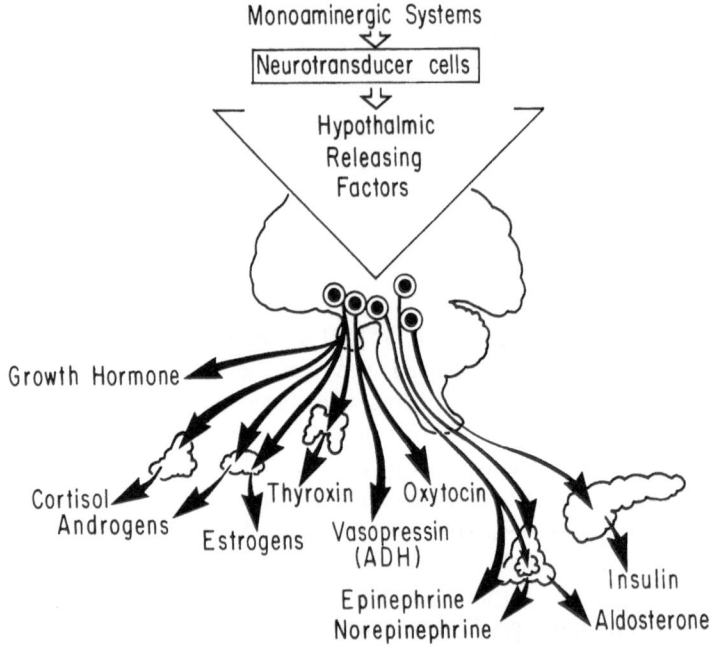

Figure 6.3 The interrelationship of the central nervous and the endocrine system.

lenge. Only later, after the novel stimulus has been mastered either by controlling the environment or by physiological adaptation, do the anabolic (i.e., growth-promoting) systems become dominant again.

Particularly fine tuning of behavior can be achieved by dual systems of control, another commonly employed regulatory mechanism. A familiar example is that of the peripheral nervous system, where sympathetic nervous stimulation speeds the heart rate while stimulation of the parasympathetic pathway slows it down. The actual rate of the heart is then a function of the prevailing tone of the two mechanisms. Neuronal networks that have essentially opposite functions, such as activity and rest or pleasure and pain, have been identified in the brains of higher animals. In addition to having a specific anatomical locus, these systems may employ different chemical messengers. We return to these particular concepts in later sections of this chapter.

Marking Time with Neurophysiology

A brief indexing of some key events in neurophysiology emphasizes just how young the discipline of neurobiology really is. Almost all of our knowledge regarding function of the central nervous system has been acquired in the last three decades.

Everybody now accepts, of course, that nervous system activity generates electrical energy. The complex physical and chemical activity occurring constantly across the neuronal membrane produces electrical potentials which, with the appropriate technology to hand, can be monitored and recorded. Such observations provide the information basic to modern neurophysiology.

In 1790 Luigi Galvani, experimenting with a "battery" made from different metals, was able to induce twitching in a frog's leg. A century later it was generally recognized that the nervous system consisted of a collection of separate units connected together by feet or *"boutons terminaux."* It was Sherrington, however, who provided the evidence for one-way transmission at these interneuronal junctions or "synapses" and recognized the reciprocal relationships between the electrical activity of the various neurons.[6] By the late 1930s Young, Hodgkin, Huxley, Eccles, and others, working largely with isolated nerve fibers taken from the giant squid, had demonstrated that these phenomena were associated with a flux of ions across the neuronal membrane.[7]

At about the same time Berger,[8] a German clinical psychiatrist, published a paper in which he described his successful recording of electrical impulses from the human scalp. He had been able to demonstrate electrical activity which changed from low-voltage desynchronized fast waves during wakefulness to a slow rhythmic recording during sleep. At first individuals were skeptical. Obviously, such patterns could have been artifacts due to muscle activity. Then in 1934 Adrian and Matthews, with an oscilloscope previously developed to record action potentials in isolated nerve, demonstrated the same phenomenon. The validity of electroencephalography (EEG) rapidly became established, and soon abnormal wave forms were described in epilepsy.[9] In 1938 Loomis recorded variability in the normal wave patterns during sleep which he called desynchronization, but it was not until 1953 that Aserinsky and Kleitman in

Chicago noted that these changes were periodic.[10] Subsequently, as the architecture of "normal" sleep EEG has been defined, the particular disturbances that develop in depression and mania have become important clinical indicators.

How were these electrical impulses transmitted from one neuron to another? While it seems a simple question now, the answer took 30 years to formulate, a period that marked the beginnings of modern neurochemistry. From Sherrington's work it was recognized that the synapse must be an important decision-making point in neuronal transmission. It was initially presumed, however, that the gap was bridged by ion discharge. Loewi, during the 1920s, presented experiments suggesting a neurochemical transmission, but it was not until the 1940s that clear evidence was provided that communication across the synapse was by chemical messenger. Even then it was another 10 years before evidence showed that the same mechanisms occurred in brain. Von Euler in Sweden, using a bioassay technique, demonstrated in 1946 that norepinephrine, a member of the catecholamine family of monoamines, was the neurotransmitter in sympathetic ganglia, the adrenal medulla, and all adrenergic nerves. Then Holtz in Germany demonstrated that normal mammalian brain also contained norepinephrine. It was initially thought that this reflected the adrenergic innervation of the cerebral blood vessels, but in 1954 Vogt showed that distribution of the norepinephrine in brain did not coincide with the density of the blood vessels. That norepinephrine was a cerebral transmitter was thus generally agreed upon less than 30 years ago.[11] Subsequently, however, extraordinary advances in technology have revolutionized both research in neurobiology and our understanding.

A Neuroanatomy of Function

Neurochemical Mapping

Many of the major neurochemical transmitters in the brain turned out to be monoamines.* Some of these molecules will fluoresce when specially prepared and viewed under light of appropriate wavelengths. Such a technique has been used for quantitative analysis and also as the backbone of several ingenious histochemical methods.[12] These have permitted the development of "maps" of the monoamine-containing nerve fibers in freeze-dried sections of the central nervous system. Two individuals, Dahlström and Fuxe in Sweden, provided the initial leadership in this field.[13] In combination with the technique of tracing degenerating axons after they have been severed, it has been possible to outline the distribution of these monoamine fibers in the brain.

The Neuroanatomic Chemistry of Emotion

The principal biogenic monoamines that are neurotransmitters in brain are 5-hydroxytryptamine (serotonin), norepinephrine, and dopamine. The highest concentra-

---

* There are two major classes of monoamines of interest to us here: the catecholamines derived from tyrosine and the indoleamines from tryptophan; both the parent substances are amino acids which occur normally in our diet.

tions of these amines are found in the midbrain (mesencephalon), the diencephalon (which connects the midbrain to the cerebral hemispheres or cortex), and the primitive brain areas of the cerebral hemispheres themselves. These are the areas of brain that have been shown through surgical ablation and neurophysiological stimulation experiments in animals to be intimately and collectively involved with such functions as the control of affect, appetite, autonomic nervous system function, aggression, and sexual behavior. These are all functions that we find disturbed in the affective disorders. Acetylcholine also is an important neurotransmitter in the control of transducer cells in the hypothalamus, cells that release the pituitary-stimulating hormones. This last amine may also have a broader role as one of the transmitters in the system of neurons that cluster around the third ventricle of the brain and seems to be involved in the expression of pain (cf. Fig. 6.4).

The amines all satisfy the criteria established for neurotransmitters. They are localized in nerve terminals, they are released when the neuron fires (depolarizes), and the enzymes necessary for their manufacture and destruction are found in the brain. It is also possible, using microneurophysiological techniques, to show that they will stimulate specific behaviors when applied to discrete brain areas.

The adrenergic amines (norepinephrine and dopamine), serotonin, and acetylcholine appear to dominate as neurotransmitters in discrete midline systems of brain stem neurons. The cells involved have extraordinarily long axons with a large number

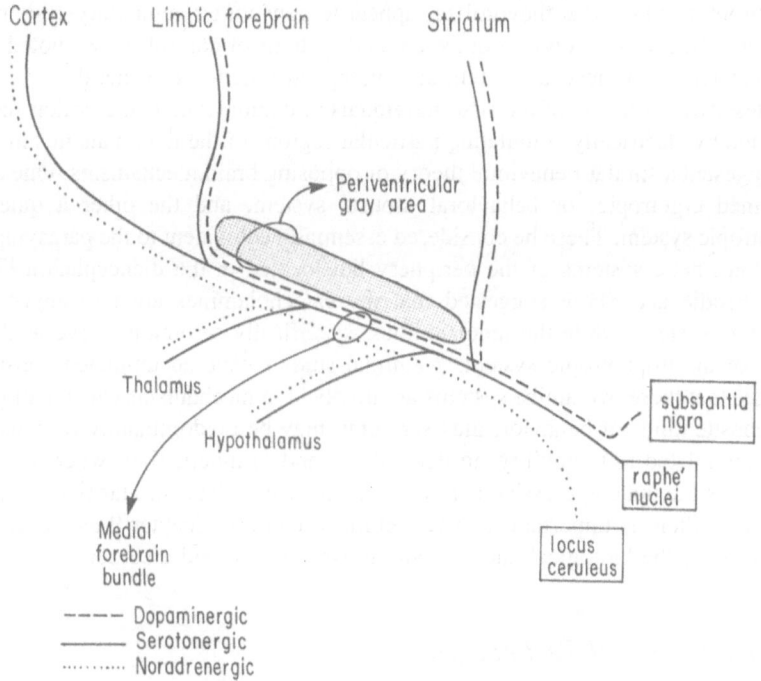

Figure 6.4 Monoaminergic pathways and the diencephalon: the diencephalon encloses the third ventricle and includes the thalamus and the hypothalamus. Stimulation along the medial forebrain bundle gives rise to reinforcement behavior and in the periventricular gray area to avoidance behavior.

of terminal branches. One neuron may have as many as 75,000 synaptic junctions; hence each cell is capable of transferring information between itself and many others. The neuronal cell bodies are primarily located in the midbrain and the diencephalon, but the axonal and dendritic projections extend to many structures found in those areas which are important in the control of emotional behavior.

These structures have been collectively termed by MacLean the visceral brain.[14] It is a functional group which includes the hypothalamus, the hippocampus, the amygdala, and certain frontal cortical areas, plus various smaller nuclei. One of the tracts of neurons which connects together these basic brain centers is the medial forebrain bundle (MFB). Stein and Wise have presented evidence that this is an adrenergic system with norepinephrine as the major neurotransmitter.[15] Olds had shown earlier that electrodes planted in nuclei and nerve fibers of the MFB cause an animal to seek stimulation repeatedly by pressing a bar to activate the electrode.[16] Because of this reward-seeking behavior, it has been suggested that this area of the MFB represents a reinforcement or pleasure center in the diencephalon. By contrast, bilateral stimulation to a cluster of neurons and fibers around the third ventricle, the periventricular system, appears to be very distressing to the animal. Most laboratory animals will perform fairly difficult tasks to avoid the stimulation. This system is thought to be serotonergically or perhaps cholinergically dominated, and because it appears to induce behaviors opposite to those of the MFB, it has been called the punishment system.[17]

These two brain mechanisms, separate in their location and probably distinct in the neurotransmitters that they utilize, appear to constitute a regulatory system which determines behavior by opposition; such dual systems of control, as we noted earlier, are characteristic of nervous system and biological function in general.

Hess, who was one of the first individuals to demonstrate that emotions could be generated by electrically stimulating particular regions of the hypothalamus in the cat, has suggested a similar behavioral theory of opposing brain mechanisms. One of these he termed ergotropic, or behavioral arousal system, and the other a quieting or trophotropic system. These he considered essentially equivalent to the parasympathetic and sympathetic systems of the periphery but located in the diencephalon.[18] In the 1950s Brodie and Shore suggested that the catecholamines are messengers of the ergotropic system, while the indoleamines, specifically serotonin, serve as the messenger of the trophotropic system.[19] Animal studies have accumulated considerable evidence that these two amine systems are involved in mechanisms capable of generating opposite kinds of behavior; thus serotonin may be predominantly responsible for behaviors related to tranquility, to deep sleep, and to hibernation, whereas the catecholamines facilitate aggressive behavior and activity.[20] The similarity of these ideas to those of Stein is important, and we return to them in Chapter 9 as we discuss an integration of the biological and behavioral aspects of mood disorders.

The Importance of Technology

Technology has been key to most of these advances. Increasingly, the techniques of the anatomist and the neurophysiologist have become combined with those of the biochemist. Frequently one advance has guided the technology of another. The electron microscope is capable of viewing the area of the synapse itself. In the presynaptic cell, dark-staining vesicles have been revealed which, using the fluorescent technique, appear to have a very high concentration of biogenic amines and are now known to be the storage sites for the neurotransmitters. Chemical methods were then developed that made it possible to disrupt the neuronal cell membrane and shake the internal contents of the neuron loose into suspension. The physical properties of the nuclei, mitochondria, bits of membrane, and other materials in this brain soup permitted them to be separated into fractions by the centrifuge (Fig. 6.5). One of these fractions contained nerve-ending particles called synaptosomes, which had within them storage vesicles, mitochondria, and some neuronal cytoplasm. In the preparation of these synaptosomes, the centrifuge was essentially used as a tool of blunt dissection, making it possible to isolate the parts of the neuron containing the highest concentration of neurotransmitter.

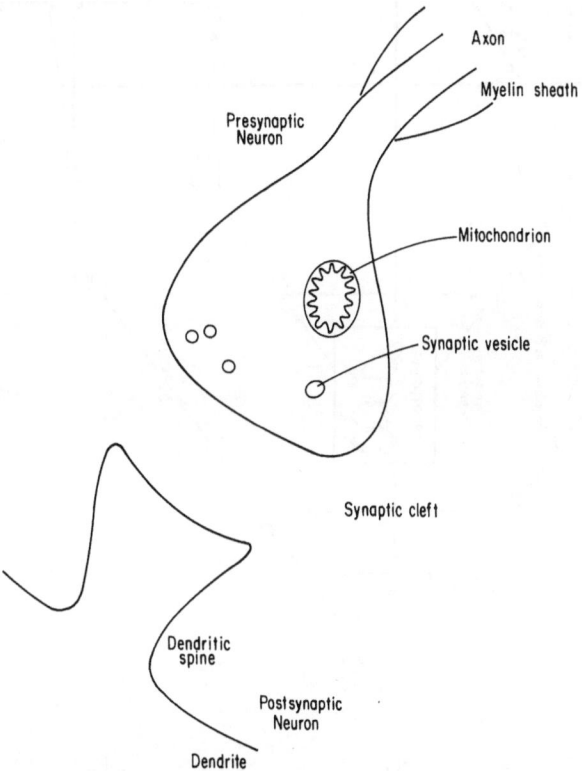

Figure 6.5 The synapse: The junction between two nerve cells, first named by Charles Sherrington in 1897, from the Greek word meaning to clasp. The synapse is the key to information processing. In a human cortex there are approximately 10^{14} synaptic contacts or 30,000 times the number of people in the world.

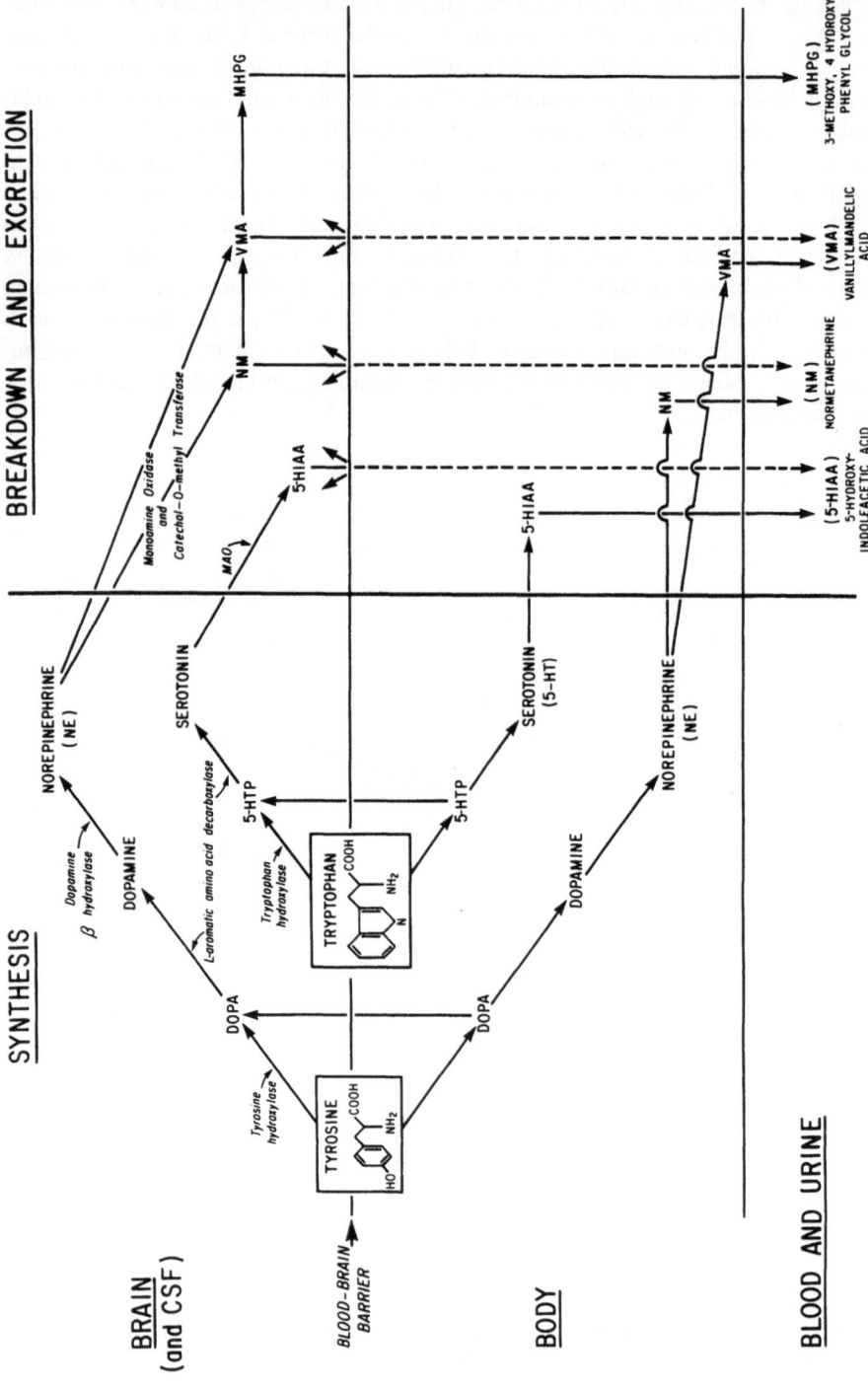

Figure 6.6 Metabolism of catechol- and indolamines: Major pathways.

This is just one of many ingenious methods that have been devised over the past 20 years to investigate brain function in vitro. The use of cerebral brain slices has been a popular tool in many biochemical investigations. Tissue culture of a variety of neural elements, brain perfusion, study of the effects of brain lesions, microdissection of various brain areas, microelectrode recording, and micropipette stimulation: all these methods have assisted in clarifying the place and mode of action of the monoaminergic transmitters.

Sensitive assays, quantitatively accurate even when measuring very small amounts of substance, have been an essential complement to such techniques. Fluorescence can be used to assay biogenic amines in solution as well as for mapping. Another important advance was chromatography, a technique that combines halogens such as fluoride with the amine, which may then be detected by a device that captures electrons. At one-and-the-same time, it allowed both quantification and separation. The use of radioisotopes has been key, both to label tracer doses of neurotransmitters and also in the development of assays based upon the principles of immunology. Most recently isotopes have been increasingly used to study receptor function; the number and affinity of the sites on the receiving cell membrane and how these change in relation to available transmitters have become important keys to understanding the dynamics of the neurotransmitters at the synaptic cleft.

The Life Cycle of the Biogenic Amines

This burgeoning technology has resulted in an impressive understanding of the chemistry of the principal neurotransmitters, norepinephrine and serotonin. These monoamines have important roles in the peripheral nervous system and the endocrine system, in addition to their roles in brain. However, beyond the initial hydroxylation of tyrosine and tryptophan, the parent amino acids, to dopa and 5-hydroxytryptophan respectively, there is a physiological barrier which prevents peripheral neurotransmitters from crossing into the brain (Fig. 6.6). As we shall see in Chapter 7, because of this barrier measurements of neurotransmitter activity in peripheral fluids do not always usefully reflect activity in the central nervous system. This has confounded much of the clinical investigation into affective illness.

Manufacture

The neuronal cell body, or nucleus, manufactures the enzymes necessary for the transformation of the raw amino acids tyrosine and tryptophan into the active neurotransmitters: norepinephrine, dopamine, and serotonin. These enzymes and the neurotransmitters themselves are actively transported down the axon to the presynaptic site, where the latter are stored prior to release.

Hydroxylation of the parent amino acids is the first and most critical step in the synthesis of both the catecholamines and serotonin. Placement of the hydroxyl group

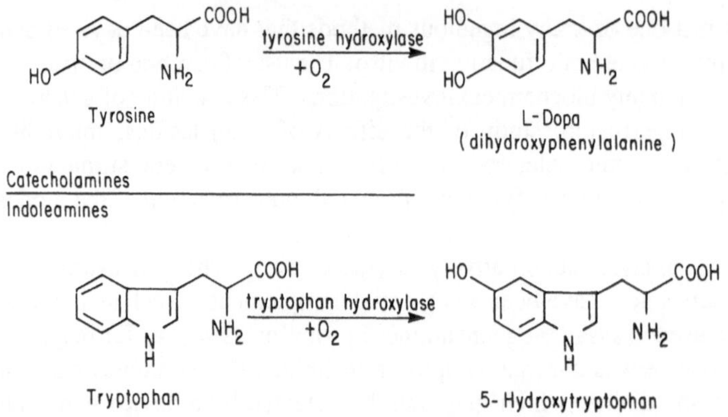

Figure 6.7 Hydroxylation in synthesis of amines. Hydroxlyation of the amino acids tyrosine and L-tryptophan is the first step in the development of the biogenic amine neurotransmitters.

on the ring structure (see Fig. 6.7) is achieved through the activity of tyrosine hydroxylase and tryptophan hydroxylase. These enzymes are found in separate neurons, corresponding to the neurotransmitter ultimately utilized. Enzyme activity is regulated from moment to moment by the amine produced, a mechanism of feedback inhibition similar to that described earlier for the endocrine system.

Next, removal of CO_2 from the amine occurs as shown in Fig. 6.8. Aromatic amino acid decarboxylase is responsible in both indole amine and catecholamine systems. The neurotransmitters dopamine and serotonin are thus derived. By an additional step in the catecholamine neurons (another hydroxylation by dopamine-beta-hydroxylase) the neurotransmitter norepinephrine is also created.

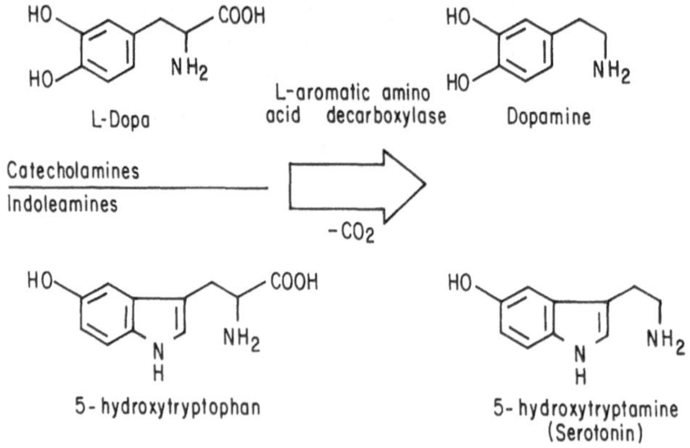

Figure 6.8 Decarboxylation in synthesis of amines. Removal of CO_2 (decarboxylation) yields the neurotransmitters dopamine and serotonin.

Events at the Synapse

There is a precise interdependence and tight regulation of all the events that support neurochemical transmission at the synapse. The neurotransmitters are stored in packages in the presynaptic area and released into the synaptic cleft when the neuronal membrane depolarizes by a calcium-dependent mechanism. Again, there is clear evidence of a regulatory system that helps control the rate of release; receptors (termed "autoreceptors") on the presynaptic neuron which are sensitive to the prevailing level of amine already present in the synaptic cleft feed back information to the neuron, thus modifying the rate of release of new neurotransmitter.

Having been launched, the messenger now seeks its specific postsynaptic receptor on the membrane of the receiving neuron. For communication to be precise between these cells, however, the neurotransmitter must be rapidly cleared from the cleft, or a crescendo of stimulation would occur. Several regulatory mechanisms help accomplish this: (1) reuptake of the neurotransmitter itself into the presynaptic neuron by an active process, (2) destruction of the neurotransmitter by enzymes, principally catechol-O-methyl transferase (COMT) within the synaptic cleft and monoamine oxidase in the mitochondria of the presynaptic nerve, (3) modulation of the further release of neurotransmitter by presynaptic receptors, and (4) changes in the postsynaptic receptor sensitivity to the neurotransmitter. The latter occur more slowly and modulate neuronal response to best utilize the amount of messenger available. A summary of these mechanisms is found in Fig. 6.9.

The postsynaptic events depend upon the recognition of the neurotransmitter substance by the receptor molecule of the postsynaptic cell membrane. Once the connection is made, metabolic events within the receiving neuron are activated by what has been termed the second messenger system. Adenylate cyclase, the enzyme that converts adenosine triphosphate (ATP) to cyclic AMP, is stimulated, and a perpetuation of the original message passed across the synaptic gap by the neurotransmitter is accomplished.

Subsequent Metabolism

Excess neurotransmitter, not taken back into the neuron for recycling, is cleared from the synaptic cleft by degrading enzymes. Many of the amine metabolites subsequently formed can be detected in the cerebrospinal fluid, blood and urine (see Fig. 6.5). 5-Hydroxy-indole-acetic acid (5-HIAA), normetanephrine (NM), and vanillylmandelic acid (VMA) are the major metabolites found both in the periphery and in the central nervous system. 3-Methoxy-4-hydroxy-phenylglycol (MHPG), another catecholamine metabolite, is unusual in its metabolism and has emerged to play a potentially important role in clinical studies. Current evidence suggests that perhaps 50% of the MHPG appearing in the urine is of central nervous system origin.[21] This distinguishes it from other indole amine and catecholamine metabolites, of which only a small proportion come from the central nervous system. It is this possibility that levels

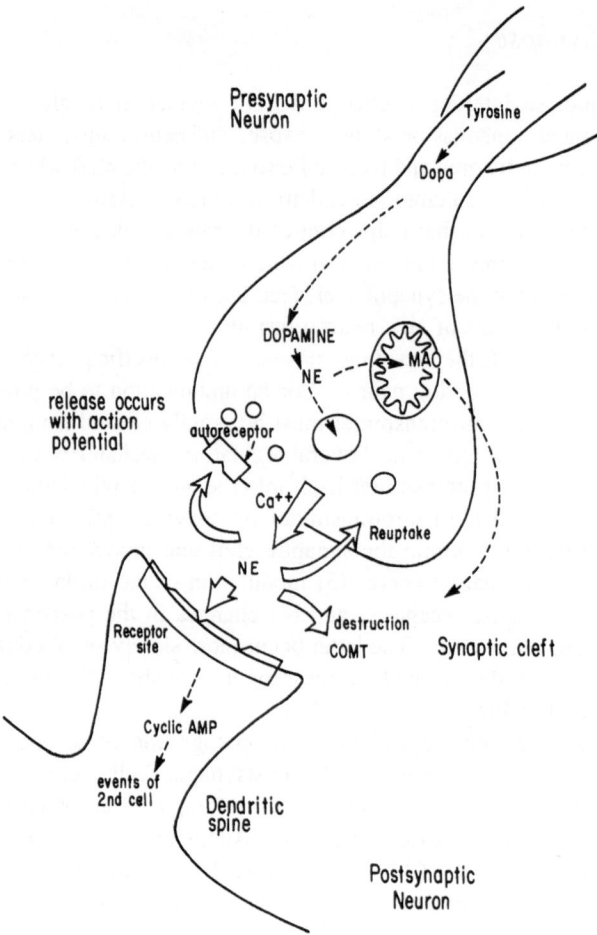

Figure 6.9 Events at the synapse.

of MHPG may reflect brain biogenic amine metabolism that has made the metabolite of particular interest to the clinician. We review some of the pertinent studies in our next chapter.

The Interface between Neurochemical and Neuroendocrine Mechanisms

We began this chapter by noting that specialized communication was the common purpose of the endocrine and nervous systems. They achieve this through very similar processes, using the principle of the chemical messenger. It is now also obvious that there are intimate connections between the two systems themselves, each influencing and directing the other in a continuous cycle of change. The predominant site of this interaction appears to be the hypothalamus, which serves as a common meeting place for the rest of the central nervous system and the pituitary—once considered the master gland of the endocrine system. Reproductive and seasonal cycles, which we now know to be driven by shifts in the daily pattern of light and dark, long ago established that there must be such a connection. However, it is only in the last decade or so that hypothalamic hormones that control the activities of the pituitary gland were identified. These pituitary-releasing factors are produced in discrete areas of the hypothalamus by cells which structurally look like neurons, with dendrites and axons. However, they have the characteristics of both endocrine and nerve cells. They manufacture and respond to endocrine-related messengers, but they also receive information, as does any neuron, through biogenic amine neurotransmitter mechanisms. Because of this capacity to translate neuronal activity into hormonal output, they have been called transducer or neurosecretory cells.

The pituitary-releasing factors that these cells produce are peptides. They are transported to the pituitary by the blood in a special portal system, and there stimulate the release of the pituitary messengers that eventually influence the peripheral endocrine glands. Earlier we reviewed the brain–thyroid system as an example of these mechanisms.

The activity of the neurosecretory cells is triggered by norepinephrine, dopamine, serotonin, and other neurotransmitters such as acetylcholine. The exact relationship between the type of neurosecretory cell, the peptide it produces, and the triggering neurotransmitter is not well understood at present, for this is a highly complex area in which to conduct research. It seems probable, however, that each neurosecretory cell manufactures a specific peptide but that the stimulation of the cell is through a combination of neurotransmitters. Hence, for example, it is probably the balance of norepinephrine, serotonin, and possibly dopamine impinging upon the receptors of the corticotropin-releasing factor (CRF) neurosecretory cell that determines whether CRF will be released. So here again, as in earlier examples, we find multiple control mechanisms combining to determine one specific outcome. However, this situation is even more complex. It also seems probable that the neurosecretory cells produce peptides that have a neurotransmitter activity localized to the brain itself. Several of these molecules, which include vasopressin, somatostatin, the enkephalins, and beta-endorphin, have been shown to influence neuroendocrine function by local action on other neurosecretory cells and possible monoamine neurotransmitters.[22] It appears that over evolutionary time new systems of communication have been superimposed upon and intertwined with those existing, so today we find multiple functions for each messenger.

The brain itself is also the target of peripheral endocrine activity. This is established by the pathological behavioral states that exist in association with endocrinopathies such as hypothyroidism, Cushing's disease, or hyperparathyroidism. Berthold's experiments with the capon were the prototype of research into this area: the hormones influence brain function in just the same way as they influence other tissue function, by modifying the transmission of information between cells. Steroid hormones, for example, clearly modulate the interaction of neuronal networks. People who take steroids as a medical treatment frequently experience euphoria first, later to be followed by depression. Exactly how the effects occur, whether by influencing the neuronal communication directly through cell membrane action or via cellular metabolism after the hormone enters and attaches to the nucleus, is not entirely clear. It is unlikely that any single mode of action for all hormones would explain the many effects which have been described.[23]

The Neuroendocrine Response to Environmental Challenge

"White with rage," "purple with anger," "trembling with emotion"—our language is peppered with such descriptive phrases. They represent a continuing memorial to an ancient awareness that in the face of extraordinary challenge, we implicitly accept that mind and body are one. Such self-knowledge long antedates the demonstration by Cannon that in the cat frightened by a barking dog there is a rise in the catecholamines produced by the adrenal gland.[24] That work, conducted over 50 years ago, marked the beginning of research on the psychophysiology of emotion and "stress."

Early Studies

In 1950 Hans Selye outlined his concept of the general adaptation syndrome in his book *The Physiology and Pathology of Exposure to Stress.*[25] He described an increase in adrenal corticosteroid production in response to a variety of noxious (stressful) stimuli. The book came at a propitious time. The Second World War had heightened general awareness to disorders induced by stressful experience, and techniques to measure corticosteroids were moving rapidly ahead. A concentrated search began to better define the psychophysiological mechanisms of stress. In 1951 Thorn and associates, studying Harvard oarsmen in the annual boat race against Yale, found that there was a marked increase of urinary 17-hydroxycorticosteroids prior to the race.[26] A series of studies by a variety of investigators rapidly followed. Anticipation of any frightening or complex task, including parachute jumping, visiting dentists, taking exams, surgical operations, and hospitalization, was found to be a potent stimulus to the adrenal cortex. Some very interesting points then began to emerge. Although situations that were novel or demanded physical exertion increased corticosteroid production, it was in those situations in which individuals were called upon to make complex decisions within certain time constraints (such as landing an aircraft on a carrier deck) that the highest levels of steroid excretion occurred.[27] Hence, in one rather amusing report, crossing a polar ice cap was found to be less stimulating to the adrenal cortex than the daily demand of being a faculty member at the Edinburgh Medical School.[28] Subsequently, studies of enlisted men in battle situations clearly demonstrated that the 24-hour urinary excretion of corticosteroids was far higher in those held responsible for the success and safety of the group than in any of the other soldiers.[29] The degree of control that the individual could exert over the situation and the personal meaning of the challenge appeared to be critical variables in determining the physiological arousal.

Individual Differences in Response

Individual differences in the psychological management of the environment appear to play a key role in physiological response. Poe, Rose, and Mason,[30] in a very careful study of army recruits, showed that successful social adaptation as measured by psychological tests had a high correlation with a minimal increase in corticosteroids under stress.

In a series of experiments spread over many years, Frankenhaeuser and her colleagues in Scandinavia have studied peripheral levels of epinephrine and norepinephrine in a wide variety of psychological test situations.[31] In the laboratory, it has been possible to identify indices of individual behavioral efficiency. Normal subjects with higher-than-average excretion rates of epinephrine during the task period performed significantly better than those who had low epinephrine excretion rates. They made significantly fewer errors and had a consistently shorter response time. A diurnal pattern also could be detected. Those individuals who preferred to work in the morning, rising early and going to bed early, tended to have a morning peak in the circadian rhythm of their epinephrine secretion which correlated with their behavioral efficiency. Evening workers similarly had higher levels of epinephrine in the evenings when they preferred to work. Both urinary epinephrine and norepinephrine secretion rose when normal subjects performed mental or heavy physical work under stressful conditions, including the threat of mild electric shock. However, when subjects were given the opportunity to learn to control their environment, there was an interesting splitting of the epinephrine and norepinephrine responses. Epinephrine progressively decreased as control over the situation was exerted, whereas norepinephrine remained high. These findings have some support from primate work by Mason and associates[32] which suggested that epinephrine output increases in situations characterized by novelty and uncertainty, but falls as familiarity occurs. Norepinephrine excretion, however, continues to be elevated despite familiarity and seems to be a correlate of any attention-demanding activity.

Some Conclusions and Comment

These studies of neuroendocrine response to environmental stressors reflect the role of the endocrine and central nervous systems as the decisive mediators between the internal and external environment. It is clear, to use T. S. Eliot's aphorism, that indeed "all cases are unique and very similar to others."[33] Even under controlled laboratory conditions, individuals have quantitatively different responses to the same stressful stimulus. However, the response changes qualitatively in a predictable way, as familiarity with the task develops. An individual diurnal pattern in performance efficiency which correlates with neuroendocrine variables is also usually present.

The importance of environmental control, however, is probably the most interesting finding. The greater the control that the individual is able to exercise over the environmental situation, the lower the neuroendocrine arousal. This emphasizes the interactional nature of the relationship between individual and environment. It is not one of stressor and respondent, but rather a closed feedback loop in which attempts are made to shape the immediate environment just as the environment shapes the individual.

The greater the structural and temporal ordering of the environmental relationship that can be achieved the more successful the adaptation. This ordering can be accomplished by direct control over the environment (which is difficult) or by self-modification dependent on the correct prediction of future environmental patterns (also difficult but less so). This concept applies at all organizational levels of living things, from the individual cell in its environment, as was first demonstrated by Bernard, to the species in the larger world. An example of the latter is the biological planning that has developed over evolutionary time to accommodate the periodicity of our planetary environment. The sleep–wake cycle is synchronized each 24 hours by the light–dark cycle, but also the basic rhythm dictated by the earth's rotation has become part of our biological template. We have an internal clock which largely determines the events of sleeping and waking. Most other metabolic events have similar circadian rhythms which oscillate relative to each other and to environmental events. We return to these special preprogrammed aspects of biological coping later, in Chapter 8.

Stressors of social and psychological origin must also be recognized as part of a larger system of interactional behavior which has profound effect on our biology. In the study of animal behavior we easily recognize challenge for food, territory, and sexual partner as part of the broader issue of hierarchical stratification of the herd through the interaction of its members. Dominance and status are settled by threat and by fighting among the contenders. In ourselves, whether we like it or not, the situation is very similar. However, as we have seen in earlier chapters, the interaction with peers and with our surroundings is made infinitely more complex, and therefore more obscure, by our unusual intelligence. The abstracting and planning capacity of the human central nervous system can add immeasurably to the range and variety of potential coping strategies. Such skills add a protective buffer between ourselves and our surroundings. To take a mundane example, we know enough to heat our houses during the winter months, thus reducing our need for total withdrawal and hibernation.

However, this ability to conceive, abstract, and plan in the psychological sphere adds a complication. We are quite capable of dreaming up dangerous situations that do not really exist within existing patterns of social interaction. We can fall short of unrealistically conceived goals and therefore judge our true ability unreasonably harshly. We can believe ourselves to be unworthy and unloved when there is little objective evidence to support the notion. Our physiological response to these symbolic situations is no different from that which would occur if there had been a real bodily assault, a demanding physical fight for dominance with another member of the herd, or the true loss of an essential nurturing parent.

The recent advances in neurobiology have brought us much closer to understanding the psychophysiological dynamics of all this. Physiological mechanisms shift to meet environmental requirements and exert control through behaviors just as patterns of psychological defense change as the best buffer is sought. Success brings reinforcement, and repetition of the behavior is likely. The final common path of such reinforcement apears to lie in the diencephalon, perhaps in the medial forebrain bundle. A particularly noxious outcome, on the other hand, fosters withdrawal, and it is unlikely that the effort will be repeated. The structural focus here appears to be somewhere in the periventricular system of the fourth ventricle. With more research, we may find that the anatomy of these mechanisms is not precisely as we believe it to be, but it seems likely that the principle will hold. It is the shifting balance between opposing physiological systems that orchestrates and ultimately determines the behavioral response.

In the studies of the peripheral nervous and endocrine systems in primates, this shifting balance can be clearly seen. When monkeys are placed in a competitive social situation, peripheral levels of epinephrine and norepinephrine rise together with steroids.[34] However, as the outcome of the competition is decided, the neuroendocrine patterns of the dominant and nondominant animals diverge significantly. In the defeated animal, corticosteroids remain high, indeed may climb higher, whereas for the victor, there is a rapid decline. Similar changes occur in the adrenosympathetic system; arousal continues in the subordinate animal who withdraws and no longer seeks to fight.[35] In females under challenge, steroids also rise but interestingly fall precipitously if they come under the protection of a dominant male.[36]

Such physiological changes when placed in the context of social behavior in groups of animals have obvious adaptive value. After the initial struggles, after the hierarchy has been established, a period of relative calm and cooperation develops between the group members. What if in one animal this psychophysiological balance is driven beyond its capacity to adapt? Chronic stressors have been shown to produce clear-cut endocrine changes. Steroid hormones climb and remain high. Growth hormone, the biogenic amines, and the sex hormones all decline. The animal withdraws from the group and exploratory behavior virtually ceases.[35] Indeed, the description begins to seem familiar, bearing some similarity to the biological disturbance that appears to be associated with the withdrawal of human melancholy. In our next chapter we turn to a review of the biology of that syndrome.

References

1. Bernard, C. *Introduction to the study of experimental medicine.* English edition translated by Green, H. C. New York: Macmillan, 1927.
2. Snyder, S. H. The opiate receptor and morphine-like peptides in the brain. *American Journal of Psychiatry,* 1978, *135,* 645–652.
3. Marx, J. L. Brain opiates in mental illness. *Science,* 1981, *214,* 1013–1015.
4. Rasmussen, H. Organization and control of endocrine systems. In R. H. Williams (Ed.), *Textbook of endocrinology.* Philadelphia: W. B. Saunders, 1974, pp. 1–30.

5. Mason, J. W. Organisation of the multiple endocrine responses to avoidance in the monkey. *Psychosomatic Medicine*, 1968, *30*, 774–790.
6. Sherrington, C. S. *The integrative action of the nervous system*. New Haven: Yale University Press, 1906.
7. Eccles, J. C. *The physiology of nerve cells*. Baltimore: Johns Hopkins Press, 1957.
8. Gibbs, F. A. Hans Berger. In W. Haymaker and F. Schiller (Eds.), *The founders of neurology*. Springfield, Illinois: Charles C. Thomas, 1970, pp. 171–175.
9. Schwab, R. S. Historical survey. In *Electroencephalography in clinical practice*. Philadelphia: W. B. Saunders, 1951, pp. 1–7.
10. Aserinsky, E., and Kleitman, N. Regularly occurring periods of eye motility and concomitant phenomena during sleep. *Science*, 1953, *118*, 273–274.
11. Von Euler, U. Discoveries of neurotransmitter agents and modulators of neuronal functions. In F. G. Worden, J. P. Swazey, and G. Adelman (Eds.), *The neurosciences: Paths of discovery*. Cambridge, Massachusetts: MIT Press, 1975, pp. 181–187.
12. Bloom, F. E. The gains in brain are mainly in the stain. In F. G. Worden, J. P. Swazey, G. Adelman (Eds.), *The neurosciences: Paths of discovery*. Cambridge, Massachusetts: MIT Press, 1975, pp. 211–227.
13. Dahlström, A., and Fuxe, K. Evidence for the existence of monoamine-containing neurons in the central nervous system. *Acta Physiologica Scandinavica*, 1964, *62* (Suppl. 232), 1–55.
14. MacLean, P. D. The hypothalamus and emotional behaviour. In W. Haymaker, E. Anderson, and W. J. Nauta (Eds.), *The hypothalamus*, Springfield, Illinois: C. C. Thomas, 1969, pp. 659–687.
15. Stein, L., and Wise, C. D. Release of NE from hypothalamus and amygdala by rewarding medial forebrain bundle stimulation and amphetamine. *Journal of Comparative and Physiological Psychology*, 1969, *67*, 189–198.
16. Olds, J. Hypothalamic substrates of reward. *Physiological Reviews*, 1962, *42*, 554–604.
17. Wise, C. D., Berger, B. D., and Stein, L. Evidence of α-noradrenergic reward receptors and serotonergic punishment receptors in the rat brain. *Biological Psychiatry*, 1973, *6*, 3–21.
18. Hess, W. R. *The functional organization of the diencephalon*. New York: Grune and Stratton, 1957.
19. Brodie, B. B., and Shore, P. A. A concept of the role of serotonin and norepinephrine as chemical mediators in the brain. *Annals of the New York Academy of Sciences*, 1957, *66*, 631–642.
20. Kiely, W. F. From symbolic stimulus to pathophysiological response: Neurophysiological mechanisms. In Z. J. Lipowski, D. R. Lipsitt, and P. C. Whybrow (Eds.), *Psychosomatic medicine: Current trends and clinical applications*. New York: Oxford University Press, 1977, pp. 206–208.
21. Maas, J. W., Hattox, S. E., Greene, N. M., and Landis, D. H. 3-methoxy-4-hydroxyphenethyleneglycol production by human brain in vivo. *Science*, 1979, *205*, 1025–1027.
22. Guillemin, R. Beta lipotropin and endorphins: Implications of current knowledge. In D. T. Krieger and J. C. Hughes (Eds.), *Neuroendocrinology*. Sunderland, Massachusetts: Sinauer Associates, 1980.
23. Whybrow, P. C., and Hurwitz, T. Psychological disturbances associated with endocrine disease and hormone therapy. In E. J. Sacher (Ed.), *Hormones, behavior and psychopathology*. New York: Raven Press, 1976.
24. Cannon, W. B. The emergency function of the adrenal medulla in pain and the major emotions. *American Journal of Physiology*, 1914, *33*, 356–372.
25. Selye, H. *The physiology and pathology of stress*. Montreal: ACTA, 1950.
26. Renold, E. A., Quigley, T. B., Kennard, H. E., and Thorn, G. W. Reaction of the adrenal cortex to physical and emotional stress in college oarsmen. *New England Journal of Medicine*, 1951, *244*, 754–757.
27. Rubin, R. T., Miller, R. G., Arthur, R. J., and Clark, B. R. Differential adrenocortical stress responses in naval aviators during aircraft carrier landing practice. *Psychological Reports*, 1970, *26*, 71–78.
28. Simpson, H. W. Field studies of human stress in polar regions. *British Medical Journal*, 1967, *1*, 530–533.
29. Bourne, P. G. Urinary 17-OHCS levels in two combat situations. In P. G. Bourne (Ed.), *The psychology and physiology of stress: With reference to special studies of the Vietnam war*. New York: Academic Press, 1969, pp. 95–103.
30. Poe, R. O., Rose, R. M., and Mason, J. W. Multiple determinants of 17-hydroxycorticosteroid excretion in recruits during basic training. *Psychosomatic Medicine*, 1970, *32*, 369–378.

31. Frankenhaeuser, M. Experimental approaches to the study of human behaviour as related to neuroendocrine functions. In L. Levi, (Ed.), *Society, Stress, and Disease.* London: Oxford University Press, 1971, pp. 22–35.
32. Mason, J. W., Brady, J. V., and Tolliver, G. A. Plasma and urinary 17-hydroxycorticosteroid responses to 72 hour avoidance sessions in the monkey. *Psychosomatic Medicine,* 1968, *30,* 608–630.
33. Eliot, T. S. *The cocktail party.* London: Faber and Faber, 1950, p. 110.
34. Candland, D. K., and Leshner, A. I. A model of agonistic behavior: Endocrine and autonomic correlates. In L. V. DiCara (Ed.), *Limbic and autonomic systems research.* New York: Plenum, 1974.
35. Leshner, A. *An introduction to behavioural endocrinology.* New York: Oxford University Press, 1978.
36. Sassenrath, E. N. Increased adrenal responsiveness related to social stress in rhesus monkeys. *Hormone Research,* 1970, *1,* 283–298.

The Emerging Neurobiology of Mood Disorder

From the particular effect of an already well-known drug on a particular mental process, the possibility exists to better recognize the true nature of the latter.

Emil Kraepelin[1]
Psychiatry (1913)

Introduction

The direction of current biological research into affective illness has grown naturally from the burgeoning of neuroscience. The rapidly advancing field of biochemical pharmacology has been a particular stimulus, and much of the theory guiding investigation both in the clinic and in the laboratory can be traced back to this source. We will place this body of knowledge at the core of this review and draw its connections with other areas of investigation. For example, the biochemical information gathered in metabolic studies of depressed persons is complemented by recent studies in clinical psychopharmacology. Together, these two bodies of data may facilitate a more precise approach to drug treatment. Similarly, specific measures of sleep disturbance and steroid metabolism appear to be highly correlated with clinical state. Such state-dependent variables may permit a biologically based diagnostic assessment, thus assisting clinical judgment and inferring the potential importance of routine neurophysiological and endocrinological testing in the clinic setting.

The Nature of the Evidence

Investigation of the biology of mood disorders has generated a complex mixture of evidence from both animal and clinical research. It is important that the limitations of that evidence be recognized. Speculation and model-building are just as important in the biological field as in the other areas that we have reviewed. However, the building blocks are subject to similar limitations as those found in primate research or phenomenology. Biological information does not necessarily provide special insight into etiology. Table 7.1 will serve to highlight some of these issues for those unfamiliar with biological investigation.

Living systems are mercurial in their response to challenge. They adapt readily, and thus an intrusion changes the nature of what we seek to observe. Clinical research into mental illness presents a particular problem in this regard. Yet, save for the primates, there are few, if any, adequate animal models of human mood disturbance.[2] Hence clinical research, the detailed investigation of individuals with depression and mania, is essential. Two drawbacks are immediately apparent in this setting. First, we cannot hope to control all the necessary variables to provide precise research results, and second, we can only sample, for biological information, a limited number of tissues. For detailed metabolic information we must turn to the laboratory, particularly

Table 7.1 Investigating the Biological Concomitants of the Affective Disorders: The Nature of the Evidence

---

I. The object: Animal studies and clinical studies

No true animal model of affective illness exists. The primate studies are closest. Clearly the reserpinized rat cannot serve well, as the model presumes an etiology for the syndrome under investigation! Clinical studies are essential to further understanding. Basic knowledge acquired from animal work can guide these studies, which in turn suggest new basic science directions. Hence each area of effort complements the other.

II. The tissue sampled

While in animal studies the brain itself can be investigated, in clinical investigations (except those postmortem) efforts are limited to measurement in peripheral tissues of metabolites thought to be reflective of CNS change. Urine, blood, and CSF are thus the main tissues sampled. Some models of the nerve cell have been utilized (e.g., the red blood cell) in an exploration of the actions of psychotropic drugs on enzyme activity and mechanisms of membrane transport.

III. Number and timing of samples

All biological systems are constantly in flux. This flux is reflected in rhythmic change of behavior and the concomitant biology, the true nature of which can only begin to be understood if the dimension of time is introduced into the conceptual framework guiding the investigation. Circadian and even annual changes in the behavior of the system must be considered in interpretation of samples measured. A single-point sample is likewise unlikely to yield very much useful information.

IV. Method of sampling

A. *Without perturbation:* Although it may be argued that the very act of sampling is a perturbation, these mechanics can be usefully distinguished from the profound challenge and subsequent adaptation induced by psychotropic drugs and ECT, for example. Sampling of physiological or biochemical variables, together with an objective record of be-

B. *With perturbation:* Sampling the response of an organism to a standard challenge is a potent method of gathering a dynamic understanding of factors underlying behavioral change and the mechanisms of adaptation. Such challenge may be at the level of the total organism, e.g., cold swimming, a mental task in a specified time frame, the striking of

---

(continued)

Table 7.1 *(Continued)*

IV. Method of Sampling *(cont.)*

A. *Without perturbation (cont.)*

havior, has formed the cornerstone of clinical investigation. The periodic nature of the affective disorders lends strength to the method, for repeated samples may easily be correlated with the naturally changing behavioral state. A further advantage is that by the patient serving as his own control, many variables of lesser interest to the investigator may be held constant.

B. *With perturbation (cont.)*

a muscle tendon, or precisely focused upon the nature of underlying mechanisms, e.g., the transmission of information across the synapse, where a series of pharmacological interventions are aimed at challenging presynaptic, synaptic, and postsynaptic events. While the nature of the sampling necessarily differs (as noted in I), the method is applicable to both animal and clinical investigation.

V. Processing the information

Machines do not necessarily confer accuracy. In the estimation of catecholamines, for example, it is important to recognize the limits of existing chemical assay methods. Until recently, many estimations depended upon fluorometric methods which become poor both in terms of specificity and sensitivity if only low levels of amine are present in the sample. Also, in the measurement of the important metabolite MHPG, various methods have been used in assay, including gas chromatography and the mass spectroscope. A recent study of paired assays using the two techniques showed no correlation in the values obtained by the two methods. Different investigators in different laboratories may not process information obtained in apparently similar circumstances in the same way, making cross-reference of data difficult.

VI. Interpreting the information

Biological complexity appears easily confused with etiology. That changes in the metabolites of biogenic amines may be defined in the affective disorders does not necessarily imply that biogenic amine change is a *more* likely explanation of the *origin* of affective disorder than is a perceived loss of self-esteem. Investigation at each level asks a slightly different question pertaining to the same problem. One must expect a slightly different answer in each case. Age and sex of patient groups, phase and duration of the illness, and time of samples in relation to circadian and circumannual events are just a few of the variables that must be considered. The nature of the evidence can only be understood when each part is taken in relation to other data available, including physiological, behavioral, and intrapsychic levels of understanding to complement a reductionistic biological perspective.

to the rat. However, here we invade the normal central nervous system, not one that is locked in melancholic withdrawal. Again, therefore, we face limitations. Cross-reference of the information gathered in the clinical setting and in the laboratory is thus vital if we are to advance our specific understanding of the biology of mood disorder. The reader should bear in mind that this interplay is a consistent feature, although sometimes implicit, of the biological research and theory that we review in this chapter.

Serendipity: New Drug Treatments and the Pharmacological Bridge

Opium and briefly camphor were used to treat manic excitement in the late 18th century, ether and chloroform in the 19th, and bromides and barbiturates in the early part of this century. Nothing worked very well. A rational pharmacotherapy of mania and melancholia has emerged, in fact, only in the past 25 years. The original circumstances that led to our present psychopharmacopeia owe much to serendipity. The discoveries themselves, however, were frequently the result of extraordinary observation.

Cade, for example, who first used lithium in the treatment of mania in 1948, had stumbled into its potential therapeutic value via the back door. He had been investigating the possibility that urea and its products acted as a toxic agent in the precipitation of mania. Lithium urate happened to be nicely soluble in water. It was during his extensive experiments in animals that he recognized the potent psychotropic agent to be the lithium ion and not urea. In the clinical setting, the response was dramatic. A 51-year-old chronically manic individual was so improved that within weeks he was able to leave the hospital where he had been resident for 5 years.[3] Similar case histories prompted an extensive research program, especially in Europe, where lithium was introduced into clinical practice in the late 1950s. Lithium and sodium ions are similar in their biological activity. Thus, in the wake of this empirical therapeutic breakthrough, there rapidly developed a detailed investigation of electrolyte metabolism in affective illness. We return to this later in the chapter. Observations during the development of other pharmacological agents stimulated similar programs of biological research. The alkaloid reserpine was isolated from *Rauwolfia serpentina* (Indian snake root) by Muller in 1952. It was soon used extensively in hypertension. Also, as a psychotherapeutic agent it competed briefly with the phenothiazines in the treatment of manic excitement. When the latter drugs were found to be much more useful both in mania and schizophrenia, reserpine was discarded from the psychiatrist's collection of empirical psychopharmacological agents. However, it was noticed that approximately 15% of the individuals receiving the drug for hypertension developed major depressive illness. When reserpine's effect upon the release of norepinephrine at the synaptic site was established, these clinical observations became a cornerstone of the hypothesis that biogenic amine metabolism might be disturbed in affective illness.[4]

The emerging theory was to find support from several other quarters. In 1951, inappropriate mood elevation was reported among some tuberculous individuals receiving iproniazid. Later in that decade, Kline successfully used the drug in the treatment of major depression.[5] These clinical observations were buttressed by new biochemical findings. Zeller and Borsky[6] and later Udenfriend[7] had both demonstrated that hydrazine compounds such as iproniazid inhibited monoamine oxidase (MAO), one of the enzymes responsible for destroying the biogenic amines at the nerve terminal.

Then in 1957 Kuhn discovered imipramine. During a search for a variant of the phenothiazine nucleus by Geigy Pharmaceuticals, imipramine was field tested in clinically disturbed individuals. While rather ineffective against the schizophrenic process, it appeared to have some mood-elevating properties.[8] Shortly thereafter, Axelrod and his associates[9] demonstrated that one of its pharmacological properties was the blockade of the reuptake of catecholamines into sympathetically innervated tissues. Again, clinical and laboratory observation had provided important complementary information. By the end of the 1950s, this fruitful collaboration between clinical and basic scientists had advanced the routine treatment of affective illness enormously. For the first time, effective therapies were available for depression (the MAO inhibitors and tricyclic antidepressants) and for mania (lithium carbonate and the phenothiazines). A series of hypotheses to explain mood disorders within the framework of disturbed biogenic amine metabolism had also been generated. It is because much of the evidence for these ideas came from psychopharmacology that this route to a better definition of the neurobiology of mood has subsequently been called the pharmacological bridge.

Amine and psychopharmacological research was not the only area of stimulated interest. Studies in electrolyte balance and neurophysiology, particularly the sleep electroencephalogram (EEG), combined with neuroendocrine research provided a broad front for advance. Improved clinical evaluation was also fostered, for objective assessment of the syndromes became necessary to conduct meaningful clinical and pharmacological research. Weaving these threads together has continued to provide a new pattern of biological information and an unprecedented opportunity to better understand the pathophysiology of mania and melancholia.

Biochemical Correlates of the Affective Syndromes

The advances in pharmacology stimulated interest in learning more about the basic metabolism of individuals with melancholia and mania. As technology advanced and new assays became available, research groups in both Europe and the United States began to explore the concomitant changes. The biochemical studies in the clinic were essentially confined to the investigation of urine and the cerebrospinal fluid (CSF) and postmortem examination of the brain. There were many frustrations and drawbacks in these investigations, which frequently generated confusing and conflicting information.

Studies of Urinary Metabolites

In 1958 Ström-Olsen and Weil-Malherbe[9] reported an increase of catecholamine excretion in the urine during mania and a decrease during depression. This was one of the early reports that, together with the information gathered from pharmacology, gave rise to the catecholamine hypothesis of the affective disorders. Various investigators, Prange,[10] Schildkraut,[11] and Bunney and Davis[12] each separately postulated that there was a functional deficit of catecholamines at the synaptic site in depression, and an excess in mania. Coppen and others in Europe proposed a similar theory based on indoleamine (serotonin) metabolism.[13,14] A flurry of studies concerning the urinary metabolites of both amine groups followed. While generally these studies confirmed that there was an increase in catecholamine excretion in mania with a comparative reduction in depression, it was unclear whether the findings differed significantly from those in control populations. Enormous individual differences in amine excretion levels confounded the research, and because the metabolites predominantly reflected the activity of the peripheral nervous system (see Fig. 6.6), little inference could be drawn about central mechanisms.[15]

Interpretation of all metabolite studies in human subjects is confounded by this same difficulty. The central nervous system is protected from the large flushes of peripheral neurotransmitter by the blood–brain barrier, and the dynamics of the two metabolic pools are not comparable. These problems can be partially avoided by following a patient through a complete cycle of illness, and when this has been successfully accomplished, there do appear greater amounts of amine metabolized during mania than in the depressed phase of bipolar illness.[9,15,16] However, that this is an artifact of physical activity remains a possibility.[17]

As a measure of serotonin metabolism, 5-hydroxyindoleacetic acid (5-HIAA) excretion was also found to be reduced in the urine of patients with depression in most (but not all) reported studies.[18] Levels appeared to be lower in unipolar than in bipolar individuals, however, an interesting reversal of the catecholamine findings. Then, in the early 1970s, enzyme studies in various animal species indicated that aldehyde reductase predominates over aldehyde dehydrogenase in the metabolism of catecholamines in brain. The opposite pertains in peripheral tissues. This suggested that the metabolites of brain norepinephrine would follow a predominantly reductive path to 3-methoxy,4-hydroxy-phenyl glycol (MHPG), whereas vanillylmandelic acid oxidation would dominate in the peripheral nervous system (Fig. 6.6). Subsequent studies by Maas and associates in Chicago demonstrated that, indeed, depending upon the species, 25%–60% of urinary MHPG has its metabolic origin in brain nor-epinephrine.[19,20] These findings afforded the opportunity to infer changes in metabolic activity of brain catecholamines in affective disorder through the index of urinary MHPG. In initial reports, urinary MHPG levels were low during serious depression in comparison both to the levels found after recovery and to those found in control populations.[20] Subsequent studies, which we shall refer to later, have indicated that not all depressives in fact have these low levels; indeed, the amine disturbance in depression appears to be heterogeneous, a finding which has become of considerable importance in pharmacotherapy.

Investigation of the Cerebrospinal Fluid

In 1966 in a controlled study Ashcroft and colleagues in Edinburgh found 5-HIAA to be low in the CSF during major depression, the severity of which correlated with the 5-HIAA concentration.[21] Others have reported similar results.[22,23] In a study at the Medical Research Council in London, Coppen and associates found the CSF concentration of 5-HIAA to be low also in mania. The levels did not improve upon recovery.[24] Denker et al. have reported a similar continuing abnormality of CSF 5-HIAA levels after remission of major depression.[22]

Most investigators agree that 5-HIAA concentrations appear to be similar in mania and severe depression.[25] Furthermore, methysergide, a serotonin antagonist, when given to manic individuals on a double-blind basis, was found to increase the manic excitement, suggesting that the reduction in available serotonin may play a role in permitting the development of mania.[26]

Because of the small quantities of fluid available, accurate measurement of the catecholamine metabolites in the CSF is difficult. However, with the refinement of assays for MHPG, several studies of this metabolite and also of homovanillic acid

(HVA) have been undertaken. While MHPG appears to be increased in mania, no clear-cut differences between depressed individuals and control populations have emerged in studies of this metabolite or HVA.[25,27]

Why should these cerebrospinal fluid data be so confusing? One possibility is the sampling of the CSF itself. First, there is a gradient of metabolites in the CSF; the spinal level at which the specimen is obtained therefore becomes an important part of the study results. This level may vary from laboratory to laboratory and from patient to patient. Second, a lumbar puncture, in contradistinction to a urine collection over 24 hours, measures the concentration of the metabolite in the CSF at the time of day in which the specimen was drawn. The circadian rhythm found in urinary MHPG excretion is presumably also present in the CSF. Hence, if the samples from the CSF are withdrawn at different times of the day for depressives and controls, then this diurnal variation might be sufficient to obscure or fabricate differences between the groups; it may be a more potent variable than that under investigation. This possibility is strengthened by there being little correlation found in many of the studies between 24-hour urine MHPG levels and CSF levels of MHPG taken during that same time period.[28]

Studies of Brains of Suicide Victims and Postmortem Material

The study of suicide brains permits direct evaluation of the biogenic amines present in the brain at the time of death. The diagnosis of melancholia is usually retrospective, however, and thus one cannot be certain that all individuals examined were suffering from the syndrome. Further, there is rapid postmortem loss of amines by metabolism. Immediately freezing the brain once it is taken from the skull and assaying the amines as quickly as possible mitigates some of these difficulties. Using this technique, two major studies were conducted in the early 1970s: one at the Medical Research Council in London[29] and one in collaboration with the National Institute of Mental Health in Bethesda.[30] In the first study, the concentration of serotonin in the pons and medulla of the depressed subjects was found to be significantly reduced when compared with that of a control group. In the second collaborative study, the brains were kept longer prior to amine assay. No significant difference between serotonin and norepinephrine and their controls was found, although 5-HIAA was significantly reduced. If any implication at all can be drawn from these studies, it is with regard to the indoleamines. Presuming the same transport efficiency of 5-HIAA out of the brain in both groups, the lowered serotonin levels in the first study and the lowered 5-HIAA levels in the second may indicate that the synthesis of serotonin is decreased in the brain stem of suicide victims.

Other studies of postmortem material have focused on the MAO enzymes. The postmortem activity of these enzymes in brain was found to correlate positively with age. Parallel studies of the enzyme activity in plasma and platelets of 122 living persons yielded a similar correlation. Because many epidemiological studies have demonstrated an increase in the prevalence of depressive illness with age, it has been suggested that MAO activity may serve as a biological marker of those at risk.[31]

Neuropsychopharmacology and Biogenic Amine Metabolism

Julius Axelrod began his investigation of the metabolism of the catecholamines in 1957. He had been stimulated in his inquiry by the hypothesis of Hoffer and Osmond that abnormal epinephrine metabolism was at the root of schizophrenia. This was the beginning of the extensive research that led to his sharing a Nobel Prize in 1970 with Ulf von Euler and Bernard Katz.[32] Although Axelrod became extensively involved in indoleamine research when his laboratory pioneered in the synthesis of melatonin and the functions of the pineal gland, it was his catecholamine work that particularly influenced research in clinical psychiatry in the United States. The development of the catecholamine hypothesis of affective disorder in the 1960s resulted directly from the new findings. By contrast, in Europe, especially in England, clinical research focused mainly upon indoleamine mechanisms and the role of serotonin. However, in subsequent years, as the pharmacological agents available to the clinician and researcher have broadened and our knowledge of the metabolic systems involved has improved, these curious variations in research focus have tended to drop away. It is now possible to perturb the behavior of both biogenic amine systems in animals and in man with a considerable range of neuropharmacological agents, and a reasonably precise understanding of mechanism has also been achieved. The site of action and the stage of the life cycle influenced by the psychotropic agents relevant to our review is found in Fig. 7.1 and Table 7.2.

It is possible to enhance or inhibit the availability of the specific amine neurotransmitters by increasing their precursor amino acid, or blocking the enzymes responsible for their manufacture. Perturbation of the usual mechanisms that sustain the dynamic events around the synapse can also be achieved in an increasing number of ways. The important elements of this new pharmacology are outlined in the paragraphs that follow, but Table 7.2 should be referred to for comprehensive reference.

Influencing the Manufacture of Biogenic Amines

Precursor Loading

In 1959 Pare and Sandler made the first attempt to reverse depressive illness by giving the parent amino acid of a putative neurotransmitter. Small doses of the norepinephrine precursor D,L-dopa were administered over a 48-hour period to three patients, but without any observed improvement.[33] Of the several studies that have followed, the most definitive was that from the National Institute of Mental Health in Bethesda, where Goodwin and his colleagues gave up to 7.2 grams per day of L-dopa, a dose equivalent to that known to produce remission in symptomatology in some parkinsonian patients. In this double-blind study of 16 patients, only four showed any clear-cut therapeutic response of their depressive symptoms. In most instances, there was psychomotor activation but no actual change in depressive symptomatology.[34]

Similar results were obtained in studies of depressed persons receiving the serotonin precursor L-tryptophan. While it has been shown to potentiate the antidepres-

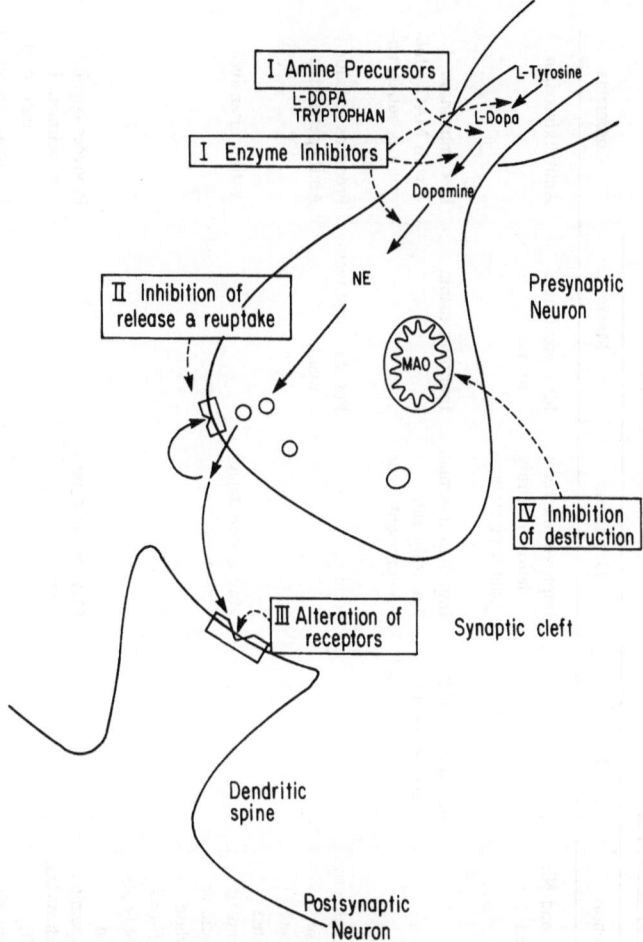

Figure 7.1 The synapse: Sites of neuronal action by pharmacological agents. See Table 7.2 for elaboration of specific drug effects.

sant effect of MAO inhibitors, no major antidepressant effect occurs when it is given alone.[35,36] However, in a double-blind trial, Coppen's group in England reported that the effects of 9 grams of L-tryptophan daily were comparable to those of imipramine in women. This advantage was not found in men.[37] The depressive symptoms that first improved were the insomnia and constipation, perhaps reflecting the somnolence and gastrointestinal symptoms that normal individuals report after taking L-tryptophan.

Thus the evidence suggests that the precursors of serotonin and norepinephrine do have behavioral effects, but these are not specifically antidepressant.

Reduction by Enzyme Inhibition

Both alpha-methyl-paratyrosine and alpha-methyl-dopa are used in the treatment of hypertension, and have been reported to induce depression. This is by no means

Table 7.2 Investigating the Biological Concomitants of the Affective Disorders: Metabolic and Behavioral Effects of Pharmacological Perturbation of Biogenic Amine Metabolism[a]

Major focus of perturbation	Drug	Presumed acute action		Behavioral effects		Comments
		Catechols	Indoles	Depression	Mania	
I. Manufacture Stimulation by precursor loading	L-Dopa	Dopamine and NE increased		Improves the retardation; not truly antidepressant	May precipitate mania	Antiparkinsonian
	L-Tryptophan		Increases 5-HTP and serotonin	Improves insomnia Not truly antidepressant alone	Improves mania	In combination with MAO inhibitor is antidepressant
Reduction by enzyme inhibition	Alpha methyl-para-tyrosine	Blocks tyrosine hydroxylase Reduces catecholamines			Possibly improves mania	Produces sedation Antihypertensive
	Alpha methyl-dopa	Blocks aromatic amino acid decarboxylase Lowers NE and presumably also serotonin		May induce depression		Antihypertensive
	Disulfiram	Blocks dopamine, beta hydroxylase Lowers NE		May induce depression		Produces cognitive impairment in some individuals
	Para-chlorophenyl-alanine	Blocks tryptophan hydroxylase Lowers serotonin				Insomnia, increased sexual behavior and aggression in animals

II. Influences the dynamics of neurotransmitter release and reuptake (presynaptic membrane effects)	Reserpine	Blocks storage of amines; hence initial increase at synapse followed by depletion	Induces depression	Produces sedation	Antihypertensive Sedating
	Amphetamine	Increases release, decreases reuptake of NE (and serotonin)	Stimulation followed by depression	May precipitate mania	Produces schizophreniform psychosis in overdose
	Tricyclic antidepressants	Depending on side chains have greater or lesser effect on catecholamines versus indoles All increase available amines by blocking reuptake	Improve depression	May precipitate mania	May improve mania if use continued through the episode
	Cocaine	Decreases reuptake NE	Poor antidepressant	May precipitate mania	Euphoriant/stimulant
	Lithium	Decreases release and storage, but also has postsynaptic effects on adenylcyclase	Antidepressant but largely prophylactic	Antimanic, prophylactic	More actions than on amine metabolism; antithyroid
	Mianserin	Blocks presynaptic alpha-adrenergic receptors, reducing the negative feedback and increasing NE release	Antidepressant		

(continued)

Table 7.2 (*Continued*)

Major focus of perturbation	Drug	Presumed acute action		Behavioral effects		Comments
		Indoles	Catechols	Depression	Mania	
III. Alters reception of neurotransmitter (postsynaptic membrane effects)	Clonidine		Presynaptic agonist decreasing NE release	Precipitates depression	Improves mania	Antihypertensive
	Salbutanol		Postsynaptic beta-2 agonist which stimulates beta-receptor sites	Rapidly improves depression		Used in asthma as bronchodilator
	Propranolol		Blocks postsynaptic beta-adrenergic sites	May precipitate depression	Improves mania	Used in thyroid storm, anxiety, and antihypertension
	Tricyclic antidepressants		Chronic administration decreases postsynaptic adrenergic receptor number and sensitivity	Improve depression	May precipitate mania	Time course for these effects similar to clinical response
	Neuroleptics, e.g., phenothiazines		Block dopaminergic (and NE) receptor sites		Improves mania	Sedation in normals
IV. Destruction Inhibition	Monoamine oxidase inhibitors		Inhibitors of monoamine oxidase increasing both NE and serotonin	Improves depression	May precipitate mania	Euphoriant in normals

[a](NE) norepinephrine.

common, however, and seems to occur predominantly in those individuals with a family history of depressive illness.[38]

Influencing the Dynamics of Release of Biogenic Amines

Once manufactured, the biogenic amines are stored pending release by depolarization of the nerve. When this occurs, through a complex process that includes calcium and other ionic shifts, the storage vesicle migrates to the inner wall of the neuronal membrane, fuses with it, and releases its contents into the synaptic junction. Here the neurotransmitter stimulates postsynaptic receptor sites. Excess amine is metabolized by the enzyme cathechol-O-methyl transferase or taken back up into the presynaptic neuron, where it may be stored or metabolized by catabolic enzymes. Many drugs, including most of the antidepressants, influence this delicate sequence of events. The mode of action is varied, however, despite a common end result—an increase in the neurotransmitter available at the receptor site.

Amphetamine and cocaine, for example, both decrease the reuptake of norepinephrine and are stimulants in normal individuals, yet they have little therapeutic value in depressed patients. In the case of amphetamine this is probably because an increased release of norepinephrine and serotonin is stimulated, in addition to the blockade of reuptake. Neuronal stores of norepinephrine are therefore ultimately depleted. Clinically, amphetamines appear to induce greater depression after an initial stimulation.

Tricyclic compounds,[39] probably the most widely used class of antidepressant, have powerful membrane effects blockading monoamine reuptake. Curiously, however, these drugs are not stimulants when given to individuals without depression; they produce only transient mild sedation. Lithium, a drug with well-defined antimanic properties, appears also to have an effect upon the storage and release of biogenic amines. In addition, it influences transfer of ions across the neuronal mechanisms and may interfere with the adenylate cyclase (energy) mechanisms in the receptor cell.[40]

In recent years new pharmacological agents with potent effects on behavior have been evaluated and introduced into the European pharmacopoeia. Analogues of these are now becoming available in the U.S. Two examples of these new classes of drugs are mianserin and clonidine. Their site of action is the presynaptic neuron. Mianserin, an antidepressant in clinical use in Europe,* blockades presynaptic alpha-adrenergic receptors, thus reducing the negative feedback from the receptor and increasing the release of norepinephrine. This is thought to be the pharmacological explanation for its antidepressant effect.[41] Conversely, clonidine, used as an antihypertensive agent, is a presynaptic receptor agonist. Stimulating the receptor, it increases the negative feedback, thus decreasing the release of the neurotransmitter. This drug has been reported to precipitate depression, and some claim that it is an antimanic agent. If substantiated, these clinical findings will complement its theoretical action at the presynaptic site.[42]

---

*A similar antidepressant, maprotiline hydrochloride, has recently been introduced into the United States.

Influencing the Receptor Site at the Postsynaptic Neuron

In their original form, the biogenic amine hypotheses suggested that mania was secondary to an excess of available neurotransmitter at the postsynaptic site. The action of the major neuroleptics, such as the phenothiazines, supports these hypotheses. The neuroleptics act by a blockade of postsynaptic reception, predominantly of dopamine; clinically they are useful agents in mania, particularly reducing motor activity. Other drugs are now emerging that also have a specific influence on the adrenergic receptor sites of the postsynaptic neuron. Salbutanol, used largely in the treatment of asthma, has been reported to produce a rapid improvement in depressive illness.[43] It appears to be a potent agonist of postsynaptic adrenergic receptors. Conversely propranolol, a beta-adrenergic blocking agent, precipitates depression, even suicide, in predisposed individuals.[44] In high doses it can also improve mania.[45]

Recent animal work has also shown that the chronic administration of tricyclic antidepressants can alter the number, function, and circadian rhythm of alpha and beta adrenoceptors and also those of dopamine and serotonin. Exactly how these pharmacodynamics of chronic administration relate to the clinical action of the tricyclics is unclear, but because therapeutic response does not usually occur for 10–14 days, they are probably very relevant. A single dose of desipramine, for example, in the rat causes a potent blockade of norepinephrine reuptake and also a weak blockade of postsynaptic alpha-1 receptors. Three weeks of treatment, however, causes an additional reduction in the number and function of beta-adrenergic postsynaptic receptors and also of the presynaptic alpha-2 receptors.[46] The latter effect would increase the release of norepinephrine by decreasing the feedback inhibition on the presynaptic neuron and thus should antagonize the effects of clonidine (see page 130). This indeed has been shown to occur in man.[47]

Decreasing the Destruction of Monoamines

Monoamine oxidase inhibitors blockade the MAO enzymes, major enzymes in the degradation of both norepinephrine and serotonin. Empirically, they seem to work in a population clinically distinct from that which responds to the tricyclic antidepressant drugs. However, they can precipitate mania in predisposed individuals,[48] and some investigators have reported that they are a euphoriant even in normal individuals.

The Evidence in Summary

What may we conclude? Taken together, the data from biochemical pharmacology support a consistent pattern of association between behavior and the amount of the available neurotransmitter at the synapse. Agents that increase available biogenic amine appear to improve symptomatology in those depressed and in some, presumably predisposed individuals, to precipitate mania. Conversely, if the drug reduces amine available at the synaptic site, then frequently it is useful in mania but may precipitate depression. Hence in general it appears that the biogenic amine hypotheses of affective illness are supported. It must be remembered, however, that the action of many of the drugs is nonspecific; they influence both catechol- and indoleamines. Also, the clinical response is not concomitant with the acute effects of the drug. Where the pharmacological effect is precise sometimes specific inference may be drawn; the precursor studies with L-dopa suggest that when norepinephrine and dopamine increase together, increased motor activity is the predominant change. Tryptophan, on the other hand, specifically increasing serotonin, improves sleep and bowel function.

Two important implications have emerged from these data. First, as pharmacological agents become increasingly specific in their biochemical effects, and assessment of metabolic function in the clinical situation improves, the hope has been generated that the therapeutic agent can be matched with the metabolic needs of the individual. The response to specific psychotropic medication would thus become increasingly predictable. Secondly, a unimodal hypothesis of biogenic amine disorder must be discarded. A view more in keeping with the information from the research is that a regulatory imbalance exists among a number of necessary biogenic amine neurotransmitters in mania and melancholia. We shall take each of these considerations in turn.

The Prediction of Pharmacological Response

A more precise understanding of the pharmacological action of antidepressants and better assays of indole- and catecholamine metabolites in the clinical setting have led to studies seeking to predict individual response to specific therapeutic agents. Most of the studies have involved the tricyclic antidepressants. Key to understanding the rationale is that the amine system particularly influenced by these drugs varies with changes in the basic tricyclic nucleus. Both of the parent or "tertiary" amines, imipramine and amitriptyline, are partially converted by the liver to the corresponding secondary amines—desmethyl-imipramine (DMI) and nortriptyline respectively. An individual receiving imipramine therefore has a combination of imipramine and DMI circulating in the blood. The rate of conversion and the percentage of secondary amine circulating when a metabolic state of equilibrium is reached vary from patient to patient, but usually, after several days, more DMI than imipramine is circulating in the blood. Desmethyl-imipramine has a more potent effect upon the norepinephrine neurons than upon serotonergic ones, and so predominantly the neuronal reuptake of catecholamines is influenced.

Amitriptyline, on the other hand, influences predominantly indoleaminergic neurons, blocking the reuptake of serotonin. Although as with imipramine the drug is demethylated (to nortriptyline), the equilibrium established between the tertiary and secondary components is different. A patient receiving amitriptyline will have a predominance of it in the blood, and the biogenic amine system most influenced is that of serotonin (see Table 7.3). The pharmacotherapist is thus theoretically afforded an opportunity to choose between agents that differentially influence the two biogenic amine systems. Amitriptyline will most benefit those with evidence of low serotonergic activity; this might be reflected by low 5-HIAA in the urine or CSF. Conversely, those with low urinary MHPG excretion as evidence of a deficient noradrenergic system would benefit from imipramine.

Table 7.3 Summary of Effects of Tricyclic Antidepressant Drugs on Blockade of Uptake of Biogenic Amines[a]

| Drug | Biogenic amine influenced | |
	Serotonin	Norepinephrine
Amitriptyline	+ + + +	+
Nortriptyline	+ +	+ +(+)?
Imipramine	+ +	+ +
Desipramine		+ + + +

[a]From reference 49.

During the 1970s several studies were conducted to test these concepts in affectively ill persons. Studies of 24-hour urinary MHPG excretion revealed that those with bipolar illness and with schizoaffective states both had lower levels of excretion of the metabolite than normals. In unipolar depression, the excretion rates were broadly distributed throughout the normal range. While there did appear to be a clustering of persons with low MHPG output, there was a significant number of depressed individuals who had high excretion. Hence the distribution of 24-hour MHPG excretion in unipolar illness appears to be bimodal. Also, in most studies low MHPG was a predictor of therapeutic response to the predominantly noradrenergic tricyclics and in one study to the tricyclic maprotiline (a specifically adrenergic antidepressant).[56] Thus the hypothesis may have some clinical validity, offering a correlation between pretreatment biological variables and drug response (Table 7.4).[50]

Table 7.4 Response to Tricyclic Antidepressants of Amine Metabolite Subgroups

| Pretreatment levels of amine metabolites | Outcome on specific tricyclic antidepressants | |
	Amitriptyline group	Imipramine group
Low 5-HIAA in CSF	Poor response to nortriptyline (Asberg et al., 1971[51])	Poor response to imipramine (Goodwin and Post, 1975[52])
Low MHPG in urine	No response to amitriptyline (Schildkraut, 1973[53]; Beckmann and Goodwin, 1975[54]) Nortriptyline gives fair response (Hollister et al., 1980[57])	Good imipramine response (Maas et al., 1972[20]) Good imipramine response in unipolar depression (Beckman and Goodwin, 1975[54]; Cobbin et al., 1979[55]; Schatzberg et al., 1982[56])
High MHPG in urine	Response to amitriptyline (Schildkraut, 1973[53]; Beckmann and Goodwin, 1975[54]; Cobbin et al., 1979[55])	No response to imipramine (Maas et al., 1972[20]; Beckmann and Goodwin, 1975[54])

Table 7.5 The Two-Type Biochemical Theory of Depression[a]

	Type A (low norepinephrine)	Type B (low serotonin)
Pretreatment MHPG	Low	Normal
Initial response to dex-troamphetamine[b]	Yes	No
Response to imipramine	Yes	No
Response to amitriptyline	No	Yes
Post-treatment MHPG	Same	Decreased

[a]After Mass[49] and Schildkraut.[53]

[b]Amphetamine is used here as a diagnostic challenge: a transient improvement in the depressive symptoms during the subsequent 24 hours is indicative of those who will benefit particularly from the antidepressants that are predominantly adrenergic in activity.

The Affective Syndromes and Hypotheses of Amine Interdependence

These findings of biochemical heterogeneity within the depressive syndrome may represent an important breakthrough in rational approaches to antidepressant treatment. Based on such findings, it has been argued that there are two separate chemical species of depression: Type A (low-norepinephrine) and Type B (low-serotonin) depression (see Table 7.5). However, an alternative explanation is that at any given moment it is the *balance* of amines available at the receptor site that determines behavior and the nature of the depressive phenomena observed.

For example, methysergide (an antiserotonin agent) makes mania worse[26]; conversely, Prange has shown that adding L-tryptophan to small doses of phenothiazine improves mania faster than does phenothiazine alone.[58] Both these results suggest that increasing the available serotonin in mania diminishes the symptomatology, and, of course, there are many examples of catecholamine-enhancing drugs that can precipitate manic phenomena (refer back to Table 7.2 for details). On the basis of these observations, Prange has proposed the "permissive biogenic amine hypothesis": A deficit in central serotonergic transmission *permits* affective disorder but is insufficient for its cause. The mood disorder itself is precipitated by changes in central catecholamine function.[58] Figure 7.2 illustrates this hypothesis, further support for which has been gathered from CSF studies. Even after recovery, depression-prone individuals have abnormally low levels of 5-HIAA, the main metabolite of serotonin, in their cerebrospinal fluid.[24] Complementing this finding, Hauri has shown that a Stage 4 sleep disturbance persists in such individuals after clinical remission.[59] The normal slow-wave sleep of Stage 4 is thought to be dependent upon the serotonergic system.

Underlying this permissive hypothesis the reader will recognize the now familiar concept of behavior regulation by neurochemical systems that act in opposition to each other. Prange's hypothesis has much in common with that of Hess who earlier pro-

Figure 7.2 The permissive hypothesis of affective illness.[58] Prange has proposed that it is the *balance* between the serotonergic and catecholaminergic systems that makes the crucial difference in determining the nature of the mood disorder. (IA) indoleamine transmission, (CA) catecholeamine transmission.

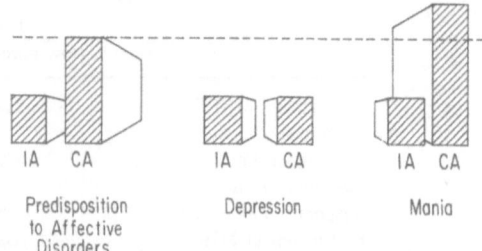

posed trophotropic (quieting) and ergotropic (arousal) mechanisms in brain based on his studies of neurophysiological stimulation in cats. The advances in neuroanatomical mapping and neurochemistry now suggest that these behavioral systems have specific neurochemical codes. Thus the trophotropic or quieting system may be coded to serotonin and/or acetylcholine and the ergotropic (arousal) system to neuroadrenergic fibers and neurotransmitters (cf. Chapter 6).

The reward system neuroanatomically located in the medial forebrain bundle (MFB) appears to be coded to norepinephrine. Stein and others have presented considerable neuropharmacological evidence for this. Chlorpromazine and reserpine, both of which reduce the effects of catecholamines, counteract the reward effects of MFB stimulation. Amphetamine, on the other hand, enhances reward behavior. Even James Olds, who first described the MFB reward system, now seems to accept this proposition.[60]

The further clarification of the anatomical distribution and chemical coding of these competing monoaminergic systems in brain promises to provide a much more specific understanding of the individual symptoms and signs present in melancholia and mania. In Table 7.6 we outline some of the tentative associations that we can draw from the limited knowledge now available.

Table 7.6 Neurophysiology–Neurochemistry–Phenomenology Correlations in Melancholia

Impaired ergotropic (reward or arousal) system: norepinephrine coded
    Emptiness, lack of emotions: "Depression beyond tears"
    Anhedonia
    Decreased social drives
    Hopeless
    Anorexia
Impaired trophotropic (quieting) system: serotonin/acetylcholine coded
    Dysphoria, pain, anxiety, agitation
    Self-reproach, guilt, low self-esteem
    Recurrent thoughts of death and suicide
Psychomotor sphere: norepinephrine, dopamine coded
    Retardation, feeling tired
    Decreased overt aggression
    Decreased initiative, poor concentration, indecisiveness

Neurophysiological Arousal States in Melancholia and Mania

Neurophysiology offers a dynamic view of the ongoing activity of neurobiological mechanisms and as such complements neurochemistry and neuropharmacology. Unfortunately, much of the psychiatric interest in neurophysiology has receded as the preoccupation with brain chemistry and pharmacology emerged. Hence a 1969 review of the existing neurophysiological studies in affective illness by Whybrow and Mendels still includes most of the major studies undertaken in the area.[61]

The late 1950s brought a particular interest in electrically recording the muscle activity (electromyography) of depressed persons. In a series of studies Whatmore described the neurophysiological state of "hyperponesis": an increased activity of pathways extending from motor and premotor cortical neurons, through pyramidal and extrapyramidal tracts, to the peripheral musculature. Investigations in both schizophrenia[62] and depression[63,64] revealed hyperponesis, but it was only in the latter that a consistent and sustained increase in muscle tension was observed. Some transient decrease occurred after electroconvulsive therapy (ECT) and improvement in the clinical state, but some degree of increased muscle activity was present in six retarded depressives even *during* clinical remission. A further rise in tension was observed just prior to the relapse of the depressive illness. In other studies Goldstein described a similar raised muscle tension in depressed outpatients as compared to normals.[65]

Extensive research of waking EEG recordings during the 1950s and 1960s revealed no specific abnormalities in affective illness. More recently, however, changes in lateral dominance have been recognized as characteristic of particular affective states. Flor-Henry from Alberta, Canada, in a series of extensive reviews, concludes that there is considerable evidence that damage to right hemisphere system (the nondominant hemisphere) is associated with depressive symptomatology. Furthermore, EEG studies in melancholia suggest a comparative shift of activity away from the normally dominant left hemisphere to the right temporoparietal region. The result is a disorganization of the usual distribution of functions between the dominant (left) and nondominant hemispheres. Greater shifts and thus greater disorganization appear to occur in mania.[66]

During EEG investigation various methods of provocation can be employed. Paired visual stimuli, for example, evoke changes of potential in the EEG in depressives that are separated by a greater period of time than in normal individuals. This prolongation of the normal cortical reactivity may also be a reflection of the greater disorganization suggested by the laterality studies.[67]

Similarly, in studying the arousal response (the blocking of the EEG alpha rhythm to an auditory or visual stimulus), Paulson and Gottlieb found that alpha suppression lasted significantly longer during depression than after recovery.[68] Another group, Wilson and Wilson, demonstrated the same phenomenon, adding that the high level of muscle tension seen in half their patients was frequently intensified by the photic stimulation.[69]

These various findings led Whybrow and Mendels to conclude that neurophysiologically the central nervous system may be considered highly aroused, although *disorganized*, during depression. One analogy is that of increased resonance. Incoming stimuli precipitate reverberating activity which is slow to settle down and thus blocks other incoming stimuli. The system then responds poorly to new information.

More recently McCarron, in an attempt to verify this hypothesis, has been able to discriminate neurophysiologically between a group of normal individuals and those experiencing a "reactive depressive syndrome." Heart rate and respiration were elevated in the depressive subjects, and in addition analysis of variance of the EEG revealed greater complexity and desynchronization. The author interpreted his findings as indicating a state of "unstable somatic activity and neurophysiological hyperexcitability."[70]

Electrolyte Studies

Extensive studies of electrolyte metabolism, particularly sodium metabolism in depression and mania, lend further support to the evidence for an increased neurophysiological arousal in these states. In 1956 Schottstaedt and associates reported a reduction in the urinary excretion of sodium during depressive episodes. Subsequently investigations of sodium imbalance in depression generally suggested a sodium retention, although in no study did this reach statistical significance.[71] Russell[72] reported it following ECT, and Anderson and Dawson[73] found an interesting tendency for depressives to decrease the amount of table salt that they consumed during depression, associated with a rising sodium retention during the illness.

Balance studies are difficult to interpret, however, and very demanding to perform for patient and researcher. Thus when isotope dilution techniques became available, which together with a total body counter made it possible to calculate changes in metabolically active sodium, there was a rapid switch to this latter technique. Coppen and Shaw, in studies that have been confirmed by most investigators, reported a significant shift of sodium into the cells during depression and mania. Shaw calculated that such a shift would decrease the average resting potential of the nerve, and concluded that the neurons of the depressive would be more excitable. The difference was of similar magnitude to the excitatory postsynaptic potential which initiates depolarization at the synapse. Hence electrolyte studies also suggest a hyperexcitable state of the CNS during depressive illness.[74]

Sleep Studies

Studies of the sleep EEG are of interest for two reasons. First, they appear compatible with the phenomena of increased arousal seen in other neurophysiological studies of depression. Second, they represent the only extensive and consistent body of data relating to disturbed circadian rhythms.

All-night EEG studies reveal several basic abnormalities of the sleep architecture in depressed people. There is a striking absence of slow-wave sleep (Stages 3 and 4), an increased difficulty in falling asleep, and an increased number of spontaneous awakenings after sleep has been achieved. There are more periods during which the level of sleep changes, and the ease of arousal from sleep by auditory stimuli is significantly increased. While in most depressives the length of sleep is actually reduced, there are some persons, particularly those who appear to have manic–depressive illness, who sleep longer during their depressive phase. Kupfer and Foster at Pittsburgh have drawn particular attention to the shortened period between sleep onset and the first dreaming period (the REM latency period). This is so consistently reduced in depressed subjects that it may be considered a state marker with potential as a biological diagnostic index.[75]

Rapid eye movement sleep (REM, dreaming) exhibits a circadian rhythm independent of the sleep–wake cycle itself. It has been suggested by Kripke and associates that in depression the circadian rhythm of this REM sleep is abnormally early and "split" from the other components of the sleep–wake cycle.[76] It is argued that severe depression represents a desynchronization of various oscillating parameters, each presumably reflecting different biological mechanisms. The diurnal temperature rhythm indeed does appear to become separated from the usual sleep–wake pattern in depression. Whereas in the normal individual the body temperature is at its nadir in the early morning, in the depressive this may occur several hours earlier. Rapid movement across geographic time zones can induce splitting of biological circadian rhythms from the sleep–wake cycle. This has led to attempts to improve depressive symptoms by forcibly changing that period in the circadian cycle during which sleep occurs.

In initial studies depressed inpatients were merely kept awake throughout the night. This appeared to have a definite therapeutic effect lasting for several days in some individuals. Furthermore, in those predisposed a switch into mania was sometimes precipitated.[77] Preliminary studies by Wehr, Goodwin, and others at the National Institute of Mental Health have now shown that the same effect can also be achieved by advancing the depressed person's sleep period by several hours. Theoretically this alters the internal phase relationship among the circadian sleep–wake cycle, the circadian REM sleep rhythm, and body temperature. Presumably if the sleep architecture was successfully returned to normal, then the mood should follow. In an initial study, going to bed 6 hours earlier than usual indeed was sufficient to return the sleep architecture to normal, with remission of symptoms.[78]

Some of these findings in interrelating the circadian rhythms of different systems may have implications for our earlier discussion of biochemical data. The disturbance among the chemically coded systems in depression may be not only one of quantitative change, such as is hypothesized by the biogenic amine theories, but also one of temporal regulation—the phase relationship of the biological mechanisms to each other along the dimension of time. Further, because regulation in the nervous system frequently is achieved through the balance of opposing mechanisms, deficiency of one amine may result in behaviors that are similar to an excess of the other. For example, in animals a deficiency of norepinephrine leads to motor inertia, decreased exploration, and apathy, while an increase in serotonin results in sedation and a diminished responsiveness to stimulation. Such behaviors may appear similar when casually observed, although the underlying biology may show considerable variation. This further emphasizes that the definition of objective biological measurements as a guide to accurate treatment and pharmacological intervention must remain an important research goal.

Changes in Endocrine Function and Physiological Mechanisms of Defense

Mental changes, and in some instances specifically mood changes, have long been recognized to occur in association with endocrine disorder. As discussed in Chapter 6, most systems are "driven" by biogenic amine mechanisms, and the hormones themselves modulate neuronal cell activity, including the activity of nervous tissue. Hence endocrine function during affective illness is potentially a very fruitful area for investigation.

The Brain–Corticosteroid Axis

The brain–corticosteroid axis has been the endocrine system most extensively studied in affective illness. Board and co-workers, in an early paper, reported elevated plasma cortisol levels at hospital admission in depressed individuals. These levels fell to normal over a 2-week period, suggesting that they were secondary to the psychological turmoil of the illness and the hospitalization.[79] Sachar, in a series of studies, confirmed this finding, but established that a subgroup of depressed patients continued to have high levels of cortisol in their blood and urine even after they had adapted to hospitalization. Clinically these individuals were described as having active suicidal thoughts, severe anxiety, and acute psychotic decompensation. Individuals with less psychological turmoil appeared to have lower levels of cortisol, but this was not a consistent finding.[80]

The subsequent development of the research in this area provides a good example of how a practical, dynamic systems approach to mental illness can clarify the evidence and improve our understanding (see Chapter 8). Early studies tended to compare serum levels of cortisol in depression with control populations of normal persons or those with other psychiatric disturbance. Subsequently serial sampling studies, following an individual through the illness and into remission, were undertaken. In bipolar illness the excretion of corticosteroids appeared to be lower during the depressed phase than in those with unipolar depression of comparable severity. However, in mania itself reports are conflicting, and only some investigations record an increasing level of cortisol as mania occurs.

It was a sophisticated series of temporal and challenge studies by Stokes[81] and Sachar et al.[82] in New York and later Carroll and associates[83] in Michigan, however, that led to our present understanding. Their pioneering work involved an assessment of the 24-hour secretion of cortisol and its pattern in relation to the sleep–wake cycle. Studies of adrenocorticotropic hormone (ACTH) production were also conducted, and it became clear that normally cortisol is secreted episodically in a series of bursts throughout the day.[84] These are synchronized with the sleep–wake cycle and thus exhibit a rhythmic oscillation. In normal individuals sleeping from 12 midnight to 8 A.M., there is very little secretion of cortisol during the late evening and early morning hours, but after about 2 A.M. a rapid rise occurs, with a peak of excretion between 5 and 9 A.M. This circadian rhythm is retained but less distinct in depression because of generally raised levels of cortisol excretion. The peak excretion also occurs earlier than in normals and is usually elevated in addition. These changes appear to tie closely with other changes found in the sleep–wake cycle itself and in the circadian REM and temperature rhythms.

Cortisol hypersecretion during depressive illness reflects hypersecretion of adrenocorticotropic hormone (ACTH), suggesting in turn that there is hyperactivity of the hypothalamic neuroendocrine centers that secrete cortisol-releasing hormone (CRH). Abnormalities in the dexamethasone suppression test support this inference. Dexamethasone is a potent synthetic corticosteroid which normally suppresses the early-morning rise in ACTH secretion by "fooling" the hypothalamus into thinking that there is sufficient cortisol already circulating. Studies by Carroll and others have shown that this suppression is lost during depression.[85] Cortisol excretion "escapes" from the dampening effect of dexamethasone during the illness but is suppressed normally after recovery from the depression. Furthermore, during the process of recovery, the dexamethasone patterns of secretion change toward normal in a predictable fashion (see Fig. 7.3). Individuals who have depressive syndromes without disturbance of biological variables such as sleep, appetite, and psychomotor function in general do not show the rapid escape characteristic of the melancholically depressed individual. Thus the dexamethasone test may be a biological variable that is dependent upon the depressed state, and some have suggested its use as a diagnostic test.[86]

Carroll concludes from these studies that in melancholia, the steroid-sensitive neurons of the hypothalamus—the steroid transducer cells—are subjected to an abnormal drive from other limbic areas. Cortisol-releasing hormone secretion is probably tonically inhibited by the noradrenergic system. Hence a reduction in hypothalamic norepinephrine, as is postulated to occur in some depressions, would release this inhibition, increasing ACTH and cortisol levels as a result.[87] Serotonin and acetylcholine are also thought to play a role, however, in the release of CRH. Thus again it is probably the balance of the control systems rather than one in particular that determines the hypothalamic drive upon the secretion of cortisol.

The possibility that elevated cortisol makes a direct contribution to the depressive

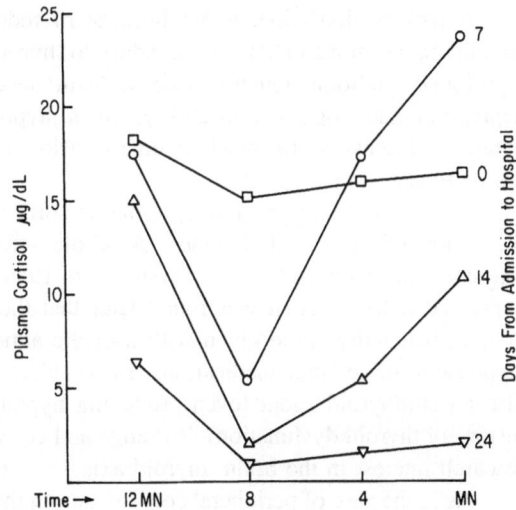

Figure 7.3 Dexamethasone suppression test profiles of a typical patient with melancholia: Note flattened and elevated profile on admission, suppression but escape after 7 days, and then a normal curve at 24 days after recovery from the depressed state.[86]

syndrome has been suggested by several researchers. One specific mechanism postulated by Curzon[88] in England and Lapin and Oxenkrug[13] in the U.S.S.R. is that raised cortisol levels, by inducing the enzyme tryptophan-pyrolase in the liver, increase the metabolism of tryptophan via the kyurenine pathway, thus causing a relative depletion of available serotonin. While attractive, this hypothesis remains largely untested.

The levels of circulating steroids in depressed persons may consistently equal those found in Cushing's disease. What is the value of such excessive excretion? One may only speculate. One is reminded of the high levels of cortisol found in the urine of primates who have withdrawn from social interaction after losing a dominance struggle. Loss and the self-assessment of having failed, so common in the depressive, can induce a similar psychobiological state of withdrawal. Perhaps we see in depression a concomitant endocrine reaction, which from the perspective of evolution had survival value, now becoming maladaptive when placed within the proximate needs of the human being.

The Brain–Thyroid Axis: An Adaptive Mechanism?

This teleological question regarding the value of physiological changes that occur in an individual developing affective disorder is brought into sharp focus when we consider the *brain–thyroid axis*. Levels of thyroxine in the blood do not vary as quickly in response to environmental challenge as do those of the 17-hydroxycorticosteroids. This is in part because thyroxine has a half-life of 8 days, but is largely secondary to the dynamics of cellular metabolism. Circulating thyroxine serves largely as a reservoir and is converted to T3 (triiodothyronine) before actually being utilized by the cells (see Fig. 6.2). This latter molecule, which has a biological half-life of some 8 hours, is four times as potent as thyroxine. Thyroid-stimulating hormone excretion is therefore an easier measure of the thyroid response to environmental challenge (e.g., temperature changes or a perceived threat), and the responsiveness of the axis is now well established.

A lack of thyroxine, either because of reduced pituitary secretion of thyroid-stimulating hormone (TSH) or secondary to thyroid gland dysfunction itself, produces a profound syndrome which includes a disturbance of mental function, predominantly depression and cognitive disability. Such hypothyroid individuals are also comparatively insensitive to catecholamine infusion; conversely, hyperthyroid individuals are more sensitive.[89]

Hence it has long been thought that in some way the thyroid molecules modulate the action of the catecholamines. A whole series of studies have now confirmed intimate links between these two systems. In 1969 Prange et al.[90] demonstrated that in depressed individuals (it was found later that these were mostly women),[91] a small dose of triiodothyronine given with tricyclic antidepressants speeded the therapeutic response to these latter medications. In combination with the evidence that lithium is also an antithyroid agent (even producing hypothyroidism in persons with a family history of thyroid dysfunction),[92] Prange and co-workers' findings have led to a rising research interest in the brain–thyroid axis.

As in the case of peripheral cortisol, serum thyroxine levels rise during depressive illness in some individuals and fall back toward normal as the depression improves. In addition recent studies of the hypothalamic mechanisms involved show that approximately 30% of depressed persons have a blunting of the usual TSH response to thyrotropin-releasing hormone (TRH) challenge during depression. This is a remarkably consistent finding across many different studies. One possible explanation is that the already raised levels of circulating thyroid are feeding back negatively upon the hypothalamus; thus a mild state of hyperthyrosis exists. Another possibility is that elevated cortisol levels interfere with the TSH response to TRH.

Whybrow has suggested that these changes in thyroid function in depression may be a mechanism of physiological defense offsetting shifts in biogenic amine metabolism, specifically catecholamine function, in the affective disorder. There is evidence of a positive correlation among the level of circulating thyroid hormone, the speed of the ankle reflex, the level of cholesterol (these latter two being dependent variables of thyroid metabolism), and the speed of response to tricyclic antidepressants.[91] Evidence is also accumulating that those with a blunted TSH response may respond more rapidly to therapeutic intervention (G. Langer, personal communication). Conversely, there appears to be a percentage of persons who develop depression who have a Grade III hypothyroid state. This is frequently unnoticed clinically but is reflected in elevated resting TSH levels and an exaggerated response to TRH challenge. In one recent survey, 9 out of 100 depressed inpatients were found to have some degree of thyroid failure.[93] Thyroid state thus appears to be of prognostic significance. The demonstration that triiodothyronine given adjunctively speeds recovery fits with this hypothesis.

More recently, evidence has developed that the mechanism of action behind these clinical observations may be an increase of beta-adrenergic receptor function at the postsynaptic site induced by increased levels of thyroid hormone. The receptor change enhances the effect of catecholamines available at the postsynaptic receptor site.[94] Even though this specific mechanism may eventually prove not to be the pertinent variable, thyroid hormones indeed are a mechanism whereby catecholamine function can be enhanced. This enhancement would presumably be of value in offsetting changes in biogenic amine metabolism that resulted in a reduction in norepinephrine occurring during depressive illness. If such adaptation is blocked, say because of raised steroid levels, or is only marginally adequate, then the individual is more prone to the development of a serious depression. The use of adjunctive thyroid hormone may then be of value.

The Role of Receptor Sensitivity in Affective Illness and the "Switch Mechanism" of Bipolar Illness

The postsynaptic neuron is capable of rapid change in its sensitivity to the available neurotransmitter. Ashcroft and colleagues,[95] Bunney et al.,[96] and others have suggested that a reduction in the amount of available neurotransmitter at the receptor site in depression, as is postulated to occur in depression, would rapidly induce an increased receptor sensitivity. Receptor supersensitivity mechanisms were first described by Cannon and Rosenblueth, secondary to denervation.[97] They may also be pharmacologically induced.*

---

* It is possible, for example, that the mechanism behind tardive dyskinesia is a supersensitivity to dopamine which develops because of a prolonged blockade by phenothiazines of the dopaminergic receptor sites.

Such receptor supersensitivity is compatible with the increased neurophysiological arousal found in depression, including the evidence from electrolyte studies. Some direct experimental evidence is also available. Ghose et al.[98] in London and Friedman[99] in the U.S., using blood pressure as a peripheral marker of central adrenergic activity, have both shown increased sensitivity to amine-enhancing pharmacological agents in depression. The dose required of both tyramine (an amine-releasing agent) and phenylephrine (a receptor agonist) to induce the same increase in blood pressure was considerably lower during depression than after recovery.

Bunney has proposed that receptor sensitivity changes may contribute to the phenomenon of "switching" from one mood state to another in manic–depressive (bipolar) cycling. These shifts can occur with considerable rapidity either spontaneously or induced by pharmacological or environmental challenge. It is suggested that during the depressive phase receptor sensitivity increases such that when pharmacological agents (e.g., amphetamine, L-dopa, or the tricyclic antidepressants) are introduced, the sudden increase of available neurotransmitter floods the receptors, triggering a massive cellular response. The associated behavioral phenomenon is a "switch" from depression to mania.[96] Pickar and associates,[100] using a similar technique to those of Ghose et al. and Friedman, have recently provided evidence to support this thesis. When blood pressure responses to tyramine were monitored during a transition from depression to mania, there was a rise in the sensitivity to the agonist just as mania developed.

Some Conclusions and Comment

The structural complexity of the central nervous system, its multiple chemical messengers, detailed regulation, and rapid response to challenge predict that manifest behavior will be a function of many variables. We are only just beginning to understand what mechanisms underpin those behaviors that are dominant in disorders of mood. While disturbed biogenic amine metabolism may represent a final common path to dysfunction, the amount of neurotransmitter in the synaptic cleft is just one of the many contributing variables. Pre- and postsynaptic regulatory mechanisms including changes in receptor sensitivity, ionic shifts across neuronal membranes, changes in enzyme activity, variations in the balance between systems regulating opposing behaviors; all are factors that we know to be relevant. Certainly there must be many more. Thus theories based upon the disturbance of a single variable are unlikely to serve us well in the future.

Consider, for example, the complexity of the adaptive mechanisms to be understood regarding the action of the antidepressants. In normally functioning brain, when a drug increases the amount of neurotransmitter available by decreasing its destruction (e.g., by inhibition of MAO), then shortly thereafter the manufacturing neuron will decrease its production of the neurotransmitter. Similarly, if neuronal reuptake is blocked, as in the administration of a tricyclic antidepressant, then postsynaptic receptors will theoretically decrease in sensitivity to accommodate the increased availability of the neurotransmitter. We know that most of the efficacious antidepressants have a rapid influence upon biogenic amine metabolism in animal experiments, and yet in normal human beings the behavioral effects are minimal. On the other hand, while undoubtedly inducing behavioral change in many depressives, clinical response to these drugs frequently does not fully develop for the first 2–3 weeks of treatment. As Emil Kraepelin recognized (see chapter quote), if these apparently disparate observations regarding drug action can be drawn together, we may learn something of importance about the mechanisms which underpin the mood disorders themselves.

Mandell suggested several years ago that the critical functions of most antidepressants were to perturb and engage the many regulatory mechanisms of the central nervous system.[101] With investigation of the chronic pharmacology of these drugs, this suggestion has assumed increasing importance. Recent research indicates that the acute increase in available neurotransmitters is indeed just the first of a series of accommodations that the neuron makes in response to the tricyclic antidepressants. Subsequently over 7–10 days, a period that closely follows the time course of clinical response, changes in receptor sensitivity occur in both the adrenergic and serotonergic systems. Presynaptic alpha-2 receptors which inhibit the release of norepinephrine from the transmitting neuron decrease in their functional sensitivity, thus presumably fostering the presence of neurotransmitter in the cleft.[47] Beta-adrenergic receptors on the postsynaptic membrane also decrease in sensitivity[102]; it is largely these receptors, when stimulated, that initiate the metabolic events that sustain the activities of the receiving neurons. Thus we may deduce that this activity is reduced, although presumably in balance. In the serotonergic system, on the other hand, changes in the sensitivity and density of serotonin receptors have not been consistent in the studies reported to date. Evidence of these adaptive mechanisms comes from the animal laboratory, but early clinical investigation suggests that the processes may be similar in depressed persons receiving desmethyl imipramine or amitriptyline—that indeed what the tricyclic antidepressants induce is a cascade of adaptive change.[46] Their principal effect is to perturb neuronal function and to foster dynamic change rather than just to increase available biogenic amines.

Pharmacological probes, including those drugs used in the clinic to alleviate depression, are also helping elucidate what appears to be a regulatory function exercised by serotonergic mechanisms upon norepinephrine neurons. Such findings have important implications for the biogenic amine hypotheses of affect.

Normetanephrine is the O-methylated metabolite of norepinephrine formed by the action of catechol-O-methyl transferase (see Chapter 6). Recently reported studies from Italy have shown normetanephrine to be a useful measure of the amount of norepinephrine being released from adrenergic systems in rat brain; in fact, the levels of normetanephrine after the acute administration of desipramine are directly related to the dose of the antidepressant administered. The studies clearly demonstrate changes in the dynamics of the system with chronic administration of the drug. After 3 days the levels of normetanephrine begin to drop even if the dose of desipramine is sustained, and after 2 weeks they are lower than in control animals. This index reflects neuronal adaptive changes that are covariant with those changes in beta-receptor-binding activity that we noted earlier. Both of these noradrenergic adaptations to tricyclic antidepressants are blunted, however, if the serotonergic system is impaired by giving 5,7,dihydroxytryptamine (a substance that destroys serotonergic neurons). The usual pattern of response can be restored by giving 5-hydroxy-tryptophan, a precursor of serotonin.[103] This suggests that serotonergic and adrenergic system are interdependent in complex ways, perhaps with the serotonergic system facilitating noradrenergic function. Sulser, using cyclic AMP as a marker, has recently demonstrated similar interconnections between the serotonergic and adrenergic systems.[104]

These basic studies in biochemical pharmacology lend support to a key concept which we discussed earlier: *that multiple mechanisms are employed by brain in the regulation of specific behaviors.* The permissive hypothesis as proposed by Prange is an example. A normally functioning serotonergic system may be necessary for the appropriate behaviors driven by the noradrenergic system to be expressed. If serotonin is diminished, then changes in the activity of norepinephrine neurons may result in aberrant behaviors such as those found in mood disorders. The highly consistent clinical findings that low levels of CSF 5-HIAA occur in depressed persons (both during illness and frequently after recovery), serotonin and 5-HIAA in the brains of suicide victims are reduced, methysergide (a general serotonin antagonist) makes mania worse, and tryptophan aids in speeding the recovery for acute mania all suggest, when considered in light of the animal studies, that the interaction between the two systems in the control of behavior is a fundamental one. Indeed, it appears certain that consideration of the dynamic interactions of brain mechanisms will become a central theme in our future understanding of the neurobiology of mood disorders. The parts can only be understood in their relationship to one another. It is this emerging opportunity that we explore in our next chapter.

References

1. Kraepelin, E. *Psychiatry,* 8th ed. Liepzig: Thieme, 1913.
2. McKinney, W. T. Animal models in psychiatry. *Perspectives in Biology and Medicine,* 1974, *17,* 529–541.
3. Cade, J. F. J. The story of lithium. In F. J. Ayd and B. Blackwell (Eds.), *Discoveries in biological psychiatry,* Philadelphia: J. B. Lippincott, 1970, pp. 218–229.
4. Goodwin, F., and Bunney, W. E., Jr. Depression following reserpine: A re-evaluation. *Seminars in Psychiatry,* 1971, *3,* 435–448.
5. Kline, N. S. Monoamine oxidase inhibitors: An unfinished picaresque tale. In F. J. Ayd and B. Blackwell (Eds), *Discoveries in biological psychiatry,* Philadelphia: J. B. Lippincott, 1970, pp. 218–229.
6. Zeller, E. A., and Borsky, J. In vivo inhibition of liver and brain monoamine oxidase by 1-iso-nicotinyl-2-isopropyl hydrazine. *Proceedings of the society for experimental biology and medicine,* 1952, *81,* 459–461.
7. Udenfriend, S. Metabolism of 5-hydroxytriptamine. In G. P. Lewis (Ed.), *5-Hydroxytriptamine: Proceedings of a symposium held in London on 1st–2nd April, 1957,* New York: Pergamon, 1958, pp. 43–49.
8. Kuhn, R. The imipramine story. In F. J. Ayd and B. Blackwell (Eds.), *Discoveries in biological psychiatry,* Philadelphia: J. B. Lippincott, 1970, pp. 205–217.
9. Ström-Olsen, R., and Weil-Malherbe, H. Humoral changes in manic–depressive psychosis with particular reference to the excretion of catecholamines in urine. *Journal of Mental Science,* 1958, *104,* 696–704.
10. Prange, A. J. The pharmacology and biochemistry of depression. *Diseases of the Nervous System,* 1964, *25,* 217–221

11. Schildkraut, J. Catecholamine hypothesis of affective disorders. *American Journal of Psychiatry*, 1965, *122*, 509–522.

12. Bunney, W. E., Jr., and Davis, M. Norepinephrine in depressive reactions. *Archives of General Psychiatry*, 1965, *13*, 483–494.

13. Lapin, I., and Oxenkrug, G. Intensification of the central serotonergic process as a possible determinant of thymoleptic effect. *Lancet*, 1969, *1*, 132–136.

14. Coppen, A. Defects in monoamine metabolism and their possible importance in the pathogenesis of depressive syndromes. *Psychiatria, Neurologia, Neurochirurgia*, 1969, *72*, 173–180.

15. Bergsman, A. The urinary excretion of adrenaline and noradenaline in some mental states. *Acta Psychiatrica et Neurologica Scandinavica*, 1959, *34*, 105–107.

16. Shinfuku, N., Omura, M., and Kayano, M. Catecholamines excretion in manic-depressive psychosis. *Yonoga Acta Medicine*, 1961, *5*, 109–114.

17. Ebert, M. W., Post, R. M., and Goodwin, F. K. Effect of physical activity on urinary 3-methoxy-4-hydroxyphenylglycol excretion in depressed patients. *Lancet*, 1972, 2, 766–767.

18. Coppen, A. Depressed states and indolealkylamines. *Advances in pharmacology*, 1968, *6*, 283–291.

19. Maas, J. W., and Landis, D. H. In vivo studies of the metabolism of norepinephrine in the central nervous system. *Journal of Pharmacology and Experimental Therapeutics*, 1968, *163*, 147–162.

20. Maas, J. W., Fawcett, J. A., and Dekirmenjian, H. Catecholamine metabolism, depressive illness and drug response. *Archives of General Psychiatry*, 1972, *26*, 252–262.

21. Ashcroft, G. W., Crawford, T. B. B., Eccleston, D., Sharman, D. F., MacDougall, F. J., Stanton, J. B., and Binns, J. K. 5-hydroxyindole compounds in the cerebrospinal fluid of patients with psychiatric or neurological diseases. *Lancet*, 1966, 2, 1049–1052.

22. Dencker, S. J., Malm, U. Roos, B., and Werdinius, B. Acid monoamine metabolites of cerebrospinal fluid in mental depression and mania. *Journal of Neurochemistry*, 1966, *13*, 1545–1548.

23. Mendels, J., Frazer, A., and Fitzgerald, R. G. Biogenic amine metabolites in cerebrospinal fluid in depressed and manic patients. *Science*, 1972, *175*, 1380–1382.

24. Coppen, A., Prange, A. J., Whybrow, P. C., and Noguera, R. Abnormalities of indoleamines in affective disorders. *Archives of General Psychiatry*, 1972, *26*, 474–478.

25. Swann, A. C., Secunda, S., Davis, J. M., Robins, E., Hanin, I., Koslow, S. H., and Maas, J. W. CSF monoamine metabolites in mania. *American Journal of Psychiatry*, 1983, *140*, 396–400.

26. Coppen, A., Prange, A. J., Whybrow, P. C., Noguera, R., and Paez, J. M. Methysergide in mania. *Lancet*, 1969, *1*, 338–340.

27. Post, R. M., Lake, C. R., Jimerson, D. C., Bunney, W. E., Wood, J. H., Ziegler, M. G., and Goodwin, F. K. Cerebrospinal fluid norepinephrine in affective illness. *American Journal of Psychiatry*, 1978, *135*, 907–912.

28. Wehr, T., and Goodwin, F. K. Catecholamines in depression. *Handbook of studies on depression*. New York: Excerpta Medica, 1977.

29. Shaw, D. M., Camps, F. E., and Eccleston, E. G. 5-hydroxytryptamine in the hindbrain of depressive suicides. *British Journal of Psychiatry*, 1967, *113*, 1407–1411.

30. Bourne, H. R., Bunney, W. E., Colburn, R. W., Davis, J. M., Davis, N. J., Shaw, D. M., and Coppen, A. Noradrenaline, 5-hydroxytryptamine and 5-hydroxyindoleacetic acid in the hindbrains of suicidal patients. *Lancet*, 1968, 2, 805–808.

31. Robinson, D. S., and Nies, A. Aging, monoamines, and monoamine oxidase levels. *Lancet*, 1972, 2, 290–291.

32. Axelrod, J. Biochemical and pharmacological approaches in the study of sympathetic nerves. In F. G. Worden, J. P. Swazey, and G. Adelman (Eds.), *The Neurosciences: Paths of Discovery*, Cambridge, Massachusetts: MIT Press, 1975, pp. 191–207.

33. Pare, C. M. B., and Sandler, M. J. A clinical and biochemical study of a trial of iproniazid in the treatment of depression. *Journal of Neurology, Neurosurgery and Psychiatry*, 1959, *22*, 247–251.

34. Goodwin, F. K., Brodie, H. K. H., Murphy, D. L., and Bunney, W. E. L-dopa, catecholamines and behavior: A clinical and biochemical study in depressed patients. *Biological Psychiatry*, 1970, *2*, 341–366.

35. Coppen, A., Shaw, D. M., Herzberg, B., and Maggs, R. Tryptophan in the treatment of depression. *Lancet*, 1967, 2, 1178–1180.

36. Herrington, R. N., Bruce, A., and Johnstone, E. C. Comparative trial of L-tryptophan and ECT in severe depressive illness. *Lancet,* 1974, *2,* 731–734.
37. Coppen, A., Whybrow, P. C., Noguera, R., Maggs, R. and Prange, A. J. The comparative antidepressant value of L-tryptophan and imipramine with and without attempted potentiation by liothyronine. *Archives of General Psychiatry,* 1972, *26,* 234–241.
38. Mendels, J., and Frazer, A. Brain biogenic amine depletion and mood. *Archives of General Psychiatry,* 1974, *30,* 447–451.
39. Gyermek, L. The pharmacology of imipramine and related antidepressants. In *International Review of neurobiology,* vol. 9. New York: Academic Press, 1966, pp. 95–143.
40. Singer, I., and Rotenberg, D. Mechanism of lithium action: Physiology in medicine. *New England Journal of Medicine,* 1973, *289,* 254–260.
41. Coppen, A., Gupta, R., Montgomery, S., Bailey, J., Burns, B., and de Ridder, J. J. Mianserin hydrochloride: A novel antidepressant. *British Journal of Psychiatry,* 1976, *129,* 342–345.
42. Charney, D. S., Heninger, G. R., Stenberg, D. E., Hafsted, K. M., Giddings, S., and Landis, D. H. Adrenergic receptor sensitivity in depression. *Archives of General Psychiatry,* 1982, *39,* 290–294.
43. Lecrubier, Y., Puech, A. J., Jouvent, R., Simon, P., and Widlocher, D. A beta-adrenergic stimulant (salbutamol) versus clomipramine in depression: A controlled study. *British Journal of Psychiatry,* 1980, *136,* 354–358.
44. Petrie, W. M., Maffucci, R. J., and Woosley, R. L. Propranolol and depression. *American Journal of Psychiatry,* 1982, *139,* 92–94.
45. Rackensperger, W., Fritsch, W., Schwarz, D., Stutte, K. H., and von Zerssen, D. Wirkung des Beta-Rezeptoren-Blockers Propanolol auf Manien. *Archiv für Psychiatrie und Nervenkrankheiten* 1976, *22,* 223–243.
46. Charney, D. S., Menkes, D. B., and Heninger, G. R. Receptor sensitivity and the mechanism of action of antidepressant treatment: Implications for the etiology and therapy of depression. *Archives of General Psychiatry,* 1981, *38,* 1160–1180.
47. Charney, D. S., Heninger, G. R., Stenberg, D. E., and Hafsted, K. M. Presynaptic adrenergic receptor sensitivity in depression: The effect of long-term desimipramine treatment. *Archives of General Psychiatry,* 1981, *38,* 1334–1340.
48. Krauthammer, C., and Klerman, G. L. Secondary mania: Manic syndromes associated with antecedent physical illness or drugs. *Archives of General Psychiatry,* 1978, *35,* 1333–1339.
49. Maas, J. W. Biogenic amines and depression: Biochemical and pharmacological separation of two types of depression. *Archives of General Psychiatry,* 1975, *32,* 1357–1361.
50. Goodwin, F. K., and Ebert, M. H. Recent advances in drug treatment of affective disorders. In M. E. Jarvik (Ed.), *Psychopharmacology in the practice of medicine,* New York: Appleton-Century-Crofts, 1977.
51. Asberg, M., Cronholm, B., Sjoquist, F., and Tuck, D. Relationship between plasma level and therapeutic effect of nortriptyline. *British Medical Journal,* 1971, *3,* 331–334.
52. Goodwin, F. K., and Post, R. M. Amine metabolites in cerebrospinal fluid, brain and urine in the major mental illnesses. In D. Freedman (Ed.), *The biology of the major psychoses: A comparative analysis,* New York: Raven Press, 1975.
53. Schildkraut, J. J. Norepinephrine metabolites as biochemical criteria for classifying depressive disorders and predicting responses to treatment: Preliminary findings. *American Journal of Psychiatry,* 1973, *130,* 695–798.
54. Beckmann, H., and Goodwin, F. K. Central norepinephrine metabolism and the prediction of antidepressant response to imipramine and amitriptyline: Studies with urinary MHPG in unipolar depressed patients. *Archives of General Psychiatry,* 1973, *32,* 17–21.
55. Cobbin, D. M., Requin-Blow, B., Williams, L. R., and Williams, W. O. Urinary MHPG levels and tricyclic antidepressant drug selection. *Archives of General Psychiatry,* 1979, *39,* 1111–1115.
56. Schatzberg, A. F., Orsulak, P. J., Rosenbaum, A. H., Maruta, T., Kruger, E. R., Cole, J. O., and Schildkraut, J. J. Toward a biochemical classification of depressive disorders. V. Heterogeneity of unipolar depressions. *American Journal of Psychiatry,* 1982, *139,* 471–475.
57. Hollister, L. E., Davis, K. L., and Berger, P. A. Subtypes of depression based on excretion of MHPG and response to nortriptyline. *Archives of General Psychiatry,* 1980, *37,* 1107–1110.
58. Prange, A. J., Wilson, I. C., Lynn, C. W., Alltop, L. B. and Stikeleather, R. A. L-tryptophan in

mania: Contribution to the permissive hypothesis of affective disorders. *Archives of General Psychiatry*, 1974, *30*, 56–62.

59. Hauri, P., Chernik, D., and Hawkins, D. Sleep of depressed patients in remission. *Archives of General Psychiatry*, 1974, *31*, 386–391.
60. Olds, J. Mapping the mind onto the brain. In F. G. Worden, J. P. Swazey, and G. Adelman (Eds.), *The neurosciences: Paths of discovery*, Cambridge, Massachusetts: MIT Press, 1975, pp. 375–400.
61. Whybrow, P. C., and Mendels, J. Towards a biology of depression: Some suggestions from neurophysiology. *American Journal of Psychiatry*, 1969, *125*, 45–54.
62. Whatmore, G. B. Some neurophysiologic differences between schizophrenia and depression, *American Journal of Psychiatry*, 1966, *123*, 712–716.
63. Whatmore, G. B., and Ellis, R. M. Some neurophysiologic aspects of depressed states: An electromyographic study. *Archives of General Psychiatry*, 1959, *1*, 70–80.
64. Whatmore, G. B., and Ellis, R. M. Further neurophysiologic aspects of depressed states: An electromyographic study. *Archives of General Psychiatry*, 1962, *6*, 243–253.
65. Goldstein, I. B. The relationship of muscle tension and autonomic activity to psychiatric disorders. *Psychosomatic Medicine*, 1965, *27*, 39–52.
66. Flor-Henry, P. On certain aspects of the localisation of the cerebral systems regulating and determining emotion. *Biological Psychiatry*, 1979, *14*, 677–698.
67. Shagass, C., and Schwartz, M. Cerebral cortical reactivity in psychotic depressions. *Archives of General Psychiatry*, 1962, *6*, 235–242.
68. Paulson, G. W., and Gottlieb, F. A longitudinal study of the electroencephalographic arousal response in depressed patients. *Journal of Nervous and Mental Disease*, 1961, *133*, 524–528.
69. Wilson, W. P., and Wilson, N. Observations on the duration of photically elicited arousal responses in depressive psychosis. *Journal of Nervous and Mental Disease*, 1961, *133*, 438–440.
70. McCarron, L. T. Psychophysiological discriminants of reactive depression. *Psychophysiology*, 1973, *10*, 223–230.
71. Schottstaedt, W. W., Grace, W. J., and Wolff, H. G. Life situations, behavior, attitudes, emotions and renal excretion of fluid and electrolytes. *Journal of Psychosomatic Research*, 1956, *1*, 287–291.
72. Russell, G. F. M. Body weight and balance of water, sodium and potassium in depressed patients given ECT. *Clinical Science*, 1960, *19*, 327–336.
73. Anderson, W. McC., and Dawson, J. Verbally retarded depression and sodium metabolism. *British Journal of Psychiatry*, 1963, *109*, 225–230.
74. Shaw, D. M. Mineral metabolism, mania and melancholia. *British Medical Journal*, 1966, *2*, 262–267.
75. Kupfer, D. J., and Foster, G. F. Interval between onset of sleep and rapid-eye-movement sleep as an indicator of depression. *Lancet*, 1972, *2*, 684–686.
76. Kripke, D. F., Mullaney, D. J., Atkinson, M., and Wolfe, S. E. Circadian rhythm disorders in manic depressives. *Biological Psychiatry*, 1978, *13*, 335–351.
77. Pflug, B. Therapeutic aspects of sleep deprivation. In *Sleep: Physiology, biochemistry, psychology, pharmacology, clinical implications*. 1st European Congress on Sleep Research, Basel, 1972. Basel: Karger, 1973, pp. 185–191.
78. Wehr, T. A., Wirz-Justice, A., and Goodwin, F. K. Phase advance of the circadian sleep–wake cycle as an antidepressant. *Science*, 1979, *206*, 710–713.
79. Board, F. A., Wadeson, R., and Persky, H. Depressive affect and endocrine function. *Archives of Neurology and Psychiatry*, 1957, *78*, 612–620.
80. Sachar, E. J., Mackenzie, J. M., Binstock, W. A., and Mack, J. E. Corticosteroid response to psychotherapy of depressions. *Archives of General Psychiatry*, 1967, *16*, 461–469.
81. Stokes, P. E. Studies in the control of adrenocortical function in depression. In T. Williams, M. Katz, and J. A. Shields (Eds.), *Recent advances in the psychobiology of depressive illness*, Washington: Government Printing Office, 1972. pp. 199–211.
82. Sachar, E. J., Hellman, I., Roffwarg, H., Halpern, F., Fukushima, D., and Gallagher, T. Disrupted 24-hour patterns of cortisol secretion in psychotic depression. *Archives of General Psychiatry*, 1973, *28*, 19–26.

83. Carroll, B. J., Curtis, G. C., and Mendels, J. Neuroendocrine regulation in depression. II. Discrimination of depressed from nondepressed patients. *Archives of General Psychiatry*, 1976, *33*, 1051–1057.

84. Hellman, L., Weitzman, E., Roffwarg, H., Fukushima, D., Yoshida, K., and Gallagher, T. F. Cortisol is secreted episodically in Cushing's syndrome. *Journal of Clinical Endocrinology and Metabolism*, 1970, *30*, 411–418.

85. Carroll, B. J., Martin, F. I. R., and Davies, B. M. Resistance to suppression by dexamethasone of plasma 11-OHCS levels in severe depressive illness. *British Medical Journal*, 1968, *3*, 285.

86. Carroll, B. J., Feinberg, M., Greden, J. F., Tarika, J., Albala, A. A., Haskett, R. F., James, N. M., Kronpol, Z., Lohr, N., Steiner, M., Vigne, J. P., and Young, E. A specific laboratory test for the diagnosis of melancholia: Standardization, validation and clinical utility. *Archives of General Psychiatry*, 1981, *38*, 15–22.

87. Sachar, E. J., Asnis, G., Nathan, S., Halbreich, U., Cabrizi, M. A., and Halpern, F. S. Dextroamphetamine and cortisol in depression. *Archives of General Psychiatry*, 1980, *37*, 755–757.

88. Curzon, G. Tryptophan pyrrolase—a biochemical factor in depressive illness? *British Journal of Psychiatry*, 1969, *115*, 1367–1374.

89. Whybrow, P. C., Prange, A. J., and Treadway, C. R. Mental changes accompanying thyroid gland dysfunction. *Archives of General Psychiatry*, 1969, *20*, 48–63.

90. Prange, A. J., Wilson, I. C., Rabon, A. M., and Lipton, M. A. Enhancement of imipramine antidepressant activity by thyroid hormone. *American Journal of Psychiatry*, 1969, *126*, 457–469.

91. Whybrow, P. C., Coppen, A., Prange, A. J., Noguera, R., and Bailey, J. E., Thyroid function and the response to L-liothyronine in depression. *Archives of General Psychiatry*, 1972, *26*, 242–245.

92. Rogers, M., and Whybrow, P. C. Clinical hypothyroidism occurring during lithium treatment: Two case histories and a review of thyroid function in 19 patients. *American Journal of Psychiatry*, 1971, *128*, 158–162.

93. Gold, M. S., Pottash, A. C., Mueller, E. A., and Extein, L. Grades of thyroid failure in 100 depressed and anergic psychiatric inpatients. *American Journal of Psychiatry*, 1981, *138*, 253–255.

94. Whybrow, P. C., and Prange, A. J. A hypothesis of thyroid–catecholamine receptor interaction: Its relevance to affective illness. *Archives of General Psychiatry*, 1981, *38*, 106–113.

95. Ashcroft, G. A., Extein, I., and Murray, L. G. Modified amine hypothesis for the etiology of affective illness. *Lancet*, 1972, *2*, 573–577.

96. Bunney, W. E., Post, R. M., and Andersen, A. E. A neuronal receptor sensitivity mechanism in affective illness: A review of evidence. *Communications in Psychopharmacology*, 1977, *1*, 393–405.

97. Cannon, W. B., and Rosenblueth, A. *The super-sensitivity of denervated structures: A law of denervation.* New York: Macmillan, 1949.

98. Ghose, K., Turner, P., and Coppen, A. Intravenous tyramine response in depression. *Lancet*, 1975, *1*, 1317–1318.

99. Friedman, M. J. Does receptor supersensitivity accompany depressive illness? *American Journal of Psychiatry*, 1978, *135*, 107–109.

100. Pickar, D., Cohen, R. M., Murphy, D. L., and Fried, D. Tyramine infusions in bipolar illness: Longitudinal changes in pressor sensitivity and behavior effects. *American Journal of Psychiatry*, 1979, *136*, 1460–1463.

101. Mandell, A. J. Neurobiological mechanisms of presynaptic metabolic adaptation and their organization: Implications for a pathophysiology of the affective disorders. In A. J. Mandell (Ed.), *Neurobiological Mechanisms of Adaptation and Behavior*, New York: Raven Press, 1975.

102. Banerjee, S. P., Kung, L. S., Riggi, S. J., and Chanda, S. K. Development of beta-adrenergic receptor subsensitivity by antidepressants. *Nature*, 1977, *268*, 455–456.

103. Racagni, G., Mocchetti, I., and Brunello, N. Presynaptic mechanisms and neurotransmitter interactions in rat central noradrenergic system after prolonged antidepressant treatment. Presented at the 21st meeting of the American College of Neuropsychopharmacology, Puerto Rico, December 15–17,

104. Janowsky, A., Okada, F., Manier, D. H., Applegate, C. D.. Sulser, F., and Steranka, L. R. Role of serotonergic input in the regulation of the beta-adrenergic receptor-coupled adenylate cyclase system. *Science*, 1982, *1982*, 900–901.

III

Toward a Synthesis

This final section brings us to an integration of those elements reviewed earlier in the book.

To integrate subjective experience and our objective understanding of biology is not a natural preference. To even begin the task demands the definition of a common paradigm. That living creatures maintain a dynamic stability through the consumption of free energy—and thus sustain a considerable degree of autonomy from their environment—is explored in Chapter 8 as such an organizing principle. Regardless of the conceptual level chosen to analyze a complex behavior, the dynamic processes relating the parts to the whole have common characteristics. These dynamic functions express a temporal morphology, patterns through time, which can provide keys to a better understanding of both normal and pathological process. Many of these patterns reflect the inherent periodicity of our planetary environment; the circadian cycle of light and dark and the seasonal changes of the annual cycle. Some of these are discussed with specific reference to the periodicity that is characteristic of many forms of affective illness.

This exploration of dynamic process serves as a prelude to Chapter 9, where we set forth the structure for an integrated psychobiology of mood disorder. In doing so we draw upon the models previously outlined in Chapter 2, emphasizing their common elements rather than their primacy. Affective illness is seen as the result of feedback interactions among three sets of interdependent variables—chemical, experiential, and behavioral—the final common path of which is mediated in the diencephalon. Predisposing elements, precipitating factors, and intermediary mechanisms are all discussed with the goal of organizing currently available information into a meaningful whole.

In Chapter 10 we extend our discussion to those implications that an integrated theory has for clinical practice, training, and research. It is suggested that organization of psychiatric practice along theoretical lines, whether biological or psychological, is no longer viable; an integrated *psychobiological* approach as exemplified by the efforts earlier in this century of Adolf Meyer is a preferred alternative. It is argued that programs providing a comprehensive approach to specific disease entities, as exemplified by disorders of mood, may aid in this transition.

Theoretical Aspects of Living Systems: Philosophical Pitfalls and Dynamic Constructs

Ivan Ilych saw that he was dying, and he was in continual despair.

In the depth of his heart he knew he was dying, but not only was he not accustomed to the thought, he simply did not and could not grasp it.

The syllogism he had learnt from Kiezewetter's Logic: "Caius is a man, men are mortal, therefore Caius is mortal," had always seemed to him correct as applied to Caius, but certainly not as applied to himself. That Caius—man in the abstract—was mortal, was perfectly correct, but he was not Caius, not an abstract man, but a creature quite, quite separate from all others. He had been little Vanya, with a mamma and a papa, with Mitya and Volodya, with the toys, a coachman and a nurse, afterwards with Katenka and with all the joys, griefs and delights of childhood, boyhood, and youth. What did Caius know of the smell of that striped leather ball Vanya had been so fond of? Had Caius kissed his mother's hand like that, and did the silk of her dress rustle so for Caius? Had he rioted like that at school when the pastry was bad? Had Caius been in love like that? Could Caius preside at a session as he did? "Caius really was mortal, and it was right for him to die; but for me, little Vanya, Ivan Ilych, with all my thoughts and emotions, it's altogether a different matter. It cannot be that I ought to die. That would be too terrible."

Leo Tolstoy[1]
The Death of Ivan Ilych (1886)

The Philosophical Pitfall of the Mind–Body Dichotomy

We come now, in this last part of the book, to a partial synthesis of the information that we have reviewed earlier. We do so with some trepidation, fully recognizing the impossibility of satisfactorily reducing the complexities involved to a comprehensive general theory.

Tolstoy's description of subjective anguish upon recognition of an untimely death illustrates one aspect of the issue. Much of the suffering in melancholia and the disorders at its fringe is highly personal. It is bound up with being alive; the intimate knowledge of the self transcends the empirical world and is little served by an intellectual understanding of psychobiology. While we enjoy far more information and sophistication in science than did Descartes, the existential dilemma of the self remains. We are each an Ivan Ilych, unique unto ourselves in life and struggling, with varying degrees of success, to develop a personal sense of order, meaning, and purpose. It is not the intent of this chapter to debate these philosophical issues; we shall simply agree with Karl Popper that the self exists as a dimension of being alive.[2]

It is our hope that the discussion in these last chapters will not automatically be interpreted as an argument for determinism; while accepting the mysteries of the self, we do not seek to perpetuate the debate over mind and body. We believe the division to be a structural pitfall, an innate human preference largely irrelevant to the broader task. It is just an easier way to think—at least until we become sick or grow old. Never as human beings have we been content to accept the self as a component of biological mechanism—or even ourselves as part of the larger animal herd. We maintain by preference an imperial posture, one reflected in our language; it contains many words useful in describing ourselves and our bodies, but rarely does any one word serve both tasks. Descartes did not invent the dichotomy of mind and body, but used it, consciously or otherwise, as an expedient which freed scientific investigation from the intrusions of the established church and the cultural dictates of his time.

Today we intellectually recognize that the integrity of mind is dependent upon the integrity of body—specifically that of brain. There is no debate of this as a fundamental postulate. Hence beyond the existential question of what it is, the dynamic qualities of mind will presumably be further illuminated by a better understanding of the fundamental processes of brain, upon which it is physically and conceptually dependent. Understanding more about the nature of psychobiological process should therefore generally assist us in thinking about ourselves. That is the essential task of this chapter, preliminary to the larger synthesis.

The approach we shall adopt is necessarily theoretical, although not exhaustively or even exclusively so. In addition to a review of dynamic theory as it relates to living organization, we shall explore in some depth biological rhythms, especially circadian and seasonal rhythms in ourselves. We shall also touch upon mathematical theory that holds promise for a future dynamic language, but sadly at present offers little more than a reminder that we must think in complex interactional terms despite our natural tendencies to reduce and divide. Finally, we shall summarize in note form those dynamic constructs that have been implicit in our earlier review and analysis of the literature in affective illness and will explicitly guide our subsequent discussion in Chapter 9.

Living Organisms as Open Systems: Dynamic Stability through Regulation and Control

Although in constant commerce with its environment, a living organism retains its autonomy. It is an open system. This characteristic is in marked distinction to a system that is closed, such as a chemical reaction in a test tube; there the interaction is governed by thermodynamic law and ultimately establishes an equilibrium defined by minimum free energy and maximum entropy (disorder or uniformity). Autonomy demands the consumption of free energy; with adequate supplies available, a living organism can survive in very varied environmental circumstances, sometimes considerably displaced from thermodynamic equilibrium.

The ordering of biological and environmental events in time and space is the key task in maintaining this independence, and as organisms grow more complicated, greater and more specialized resources are committed to this regulatory activity. A centralized nervous system may be viewed as the ultimate regulator, the role of which is to maintain all events occurring within the body in a functional sequence and sustain an orderly interaction with the environment in space and time. Claude Bernard was one of the first to recognize the need for regulation in biological systems and conducted some of the pioneering experiments to demonstrate the mechanisms of control that the body employs.[3] In Chapter 6 we reviewed the homeostatic organization of the endocrine system as an example of these regulatory devices. Similar mechanisms are found governing functions within the cells themselves; indeed, much of modern biological research is concerned with the elucidation and study of these cellular control systems.

Analogy with the Mechanical World

The definition of a living creature as a finite, energy-consuming, precisely regulated hierarchy of biological systems that, through adaptation, maintain a dynamic steady state in defiance of thermodynamic law, is relatively new. Conceptually it draws heavily upon theoretical physics and engineering, supplemented by some practical analogy with modern computer technology.

Engineers and biologists have always shared a common interest in behavior. Engineers for pragmatic reasons particularly need to determine the behavior of a whole system in relation to its component parts. Also, like biological systems, machines consume free energy and are able to do work only while their energy supply is maintained. If this fails, then they obey thermodynamic laws, as does the living organism after death. Hence, the experience gained in building machines has frequently provided useful analogies for thinking about ourselves.

Essential to Harvey's assertion that the blood circulated was his conception of the heart as a pump—a familiar enough mechanical device, even in the 1600s. Descartes, delighted with the water-animated grottoes of the royal gardens of the Renaissance, likened the "nerves of [the body] with the tubes of the machines of these fountains. . . and the animal spirits, to the . . . water that puts them in motion."[4]

The clock was another object of human ingenuity which fascinated those who, in the early 17th century, sought to describe human behavior. Following the invention of the deadbeat escapement by Graham, it rapidly became a symbol of precision, regulation, and reliability. Thus Descartes' analogy of mind and body was drawn to two clocks standing side by side.

The modern equivalent to Descartes' clock is the computer: a machine, as J. Z.

Young, the British neuroanatomist, has emphasized, that can make choices.[5] This is a fundamental step toward thinking. We have become quite comfortable with the idea that machines can do work, just as do human muscles, but there has been greater reluctance to acknowledge that technology can now create devices that may think. Young argues that just as understanding the function of the heart came from treating it as possessing the qualities of a simple pump, so can the organization of the nervous system be better understood by considering the brain as the temporal integrator of a large set of self-regulating mechanisms and hierarchically ordered homeostats—basically a computer with well-developed clocklike functions. In fact, some very interesting insights have emerged from pursuing this particular analogy.

Rhythmic Behavior and Temporal Organization in Biology

Life may be conceptualized as the perpetuation of a state of stable dynamic organization which facilitates adaptive performance. Without this stability the necessary tasks of development, learning, and management of the environment will be impeded and the life of the organism potentially compromised. The apparently stable functions of the body, for example body temperature, blood sugar, wakefulness (and even such concepts as self-esteem), however, are founded on the paradox of continuous, but regulated, change. In fact, upon temporal inspection this flux is easily observed as a rhythmic patterning of biological and behavioral variables. Such rhythms are ubiquitous in biological systems; they have have been identified in all organisms studied, from unicellular algae to man, emphasizing that they are indeed a fundamental property of life.[6,7] They are frequently overlooked in ourselves, but as noted in earlier chapters may be detected in both health and illness.

When monitored over time, homeostatically controlled systems with negative feedback demonstrate rhythmic behavior. The brain–thyroid system was the example we used earlier. The behavior of such a system describes a recurring pattern, a temporal morphology (Winfree[8]) which varies as information is received and the output appropriately adjusted.

When the decay of utilization of the output in such a mechanism is linear, and the system is not disturbed, the pattern will be a sine wave. Actually very few biological systems are so simple, and thus linear oscillation is rarely encountered. Most of the rhythms detected in nature are produced by nonlinear oscillators. The waveform of these nonlinear phenomena is still lawful but much more complex. Many patterns occur; the temporal morphology may describe a steady rise in activity followed by relaxation (e.g., the contraction of a heart ventricle) or exhibit a complex cycle of behavior which, while it does eventually reenter upon itself, the uneducated observer would find difficult to identify as rhythmic. The oscillators that produce these latter, more complex, nonlinear rhythms are known as limit-cycle oscillators. They abound in biological systems. Because of their complexity they rarely set up resonance, but they do have the important property of entrainment. This means that they tend to synchronize with a periodic driving stimulus. An example is the entrainment of biological processes by photic stimuli in the naturally occurring light–dark cycle—the rhythm of sleeping and waking or the control of the rhythm of melatonin excretion by the pineal gland.

For most biological rhythms, it is not clear what initiates these oscillations. Many occur independently of the adaptive demand of the environment. Some argue that each rhythm is tied back to a homeostatic mechanism, but little is really known about the physiochemical basis of oscillation. What the rhythms reflect, however, in dynamic terms is the ongoing behavior of a set of regulated systems which are frequently interdependent and hierarchically organized. Such rhythmic functions appear to contribute to psychobiological stability in at least two ways:

1. By placing temporal constraints upon events that if they occurred out of sequence or simultaneously would defeat their original biological purpose (e.g., cell division in the growing organism or the events of the menstrual cycle).
2. By acting as a biological planner which facilitates adaptation to a regularly changing (periodic) environment (e.g., the circadian sleep–wake cycle in response to light and dark, or annual hibernation at those latitudes of the planet where winter must be accommodated).

Such planning, which offers great adaptive advantage, requires that the organism develop a four-dimensional model or space–time template of environmental events, so that their proper sequence in space–time can be anticipated during the next cycle. During evolution such templates appear to have been incorporated into the functional structure of many organisms, and the rhythmic patterns of behavior are expressed even in the absence of the original environmental demand. Such internalized mechanisms have been labeled biological clocks, the most visible of which are those related to the circadian cycle. Many metabolic parameters when sequentially measured exhibit circadian rhythmicity, and some, such as corticosteroids, are notably disturbed in affective illness.

Human Biological Clocks

Clock metaphors pervade the literature on biological rhythms. The term is used largely for those rhythms that are locked into the environmental changes resulting from the periodicity of the planet on which we live—particularly the cycles related to the tides (12 hours, largely marine creatures) and the sun (24 hours, or circadian, and the annual cycles). Modern technical ingenuity has allowed us largely to forget our planet's periodicity. In earlier times the average person was much more aware of daily and seasonal patterns produced by our changing geophysical circumstances. Those living in a northern latitude had noticed, of course, that the sun retreated in the latter part of the year toward the south, and numerous rituals were developed in all cultures to stop that retreat and bring the sun back. Such was the true significance of the celebration of the winter solstice, which in our calendar falls in the third week of December.

Nonetheless, despite this recent intellectual neglect clock mechanisms persist and pervade man's biology. The sleep–wake cycle, a circadian rhythm, is the most obvious. It is important to recognize, of course, that these clocks are not exactly like the mechanical devices with which we tell time. While they repeat themselves every 24 hours, their behavior is nonlinear. The degree of change in unit time may progress more quickly or more slowly depending upon the phase of the cycle and the environ-

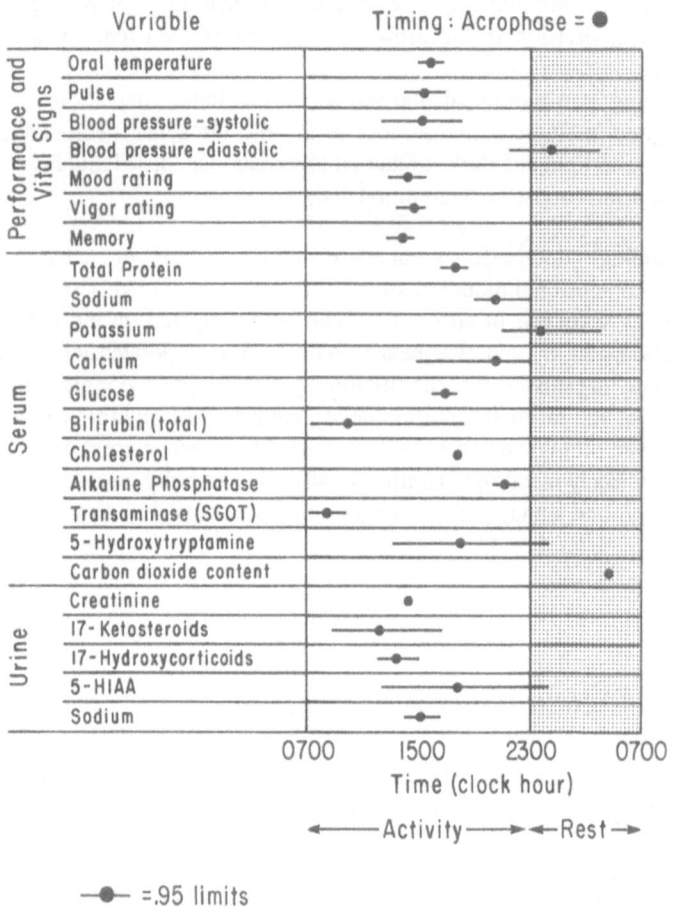

Figure 8.1 The peak of circadian variation in human physiological systems. The line through each point represents the 95% confidence regarding the variance. Hence temperature varies little in its acrophase (peak), whereas serum serotonin varies widely. (Adapted from data gathered by the Chronobiology Laboratory, University of Minnesota; Franz Hallberg[9].)

mental challenge that is encountered. As Winfree has colorfully described it, "they lurch and pause around a common cycle."[8] Most of the clocklike systems, however, do have a preferred cycle, to which they return if not disturbed by external events; they then exhibit a stable periodic pattern. Many of the clocks are interconnected or entrained to a dominant cycle, and the degree to which they can be disturbed from their preferred circadian oscillation varies. We know that in ourselves some are much more tightly controlled than others. Body temperature, for instance, despite daily challenge fluctuates very little in amplitude (no more than $1.5°$ to $2°$ F) and hardly deviates from its 24-hour period even when the individual is living without the usual time cues of the day–night cycle. Other systems are more easily perturbed and have greater variance over the circadian cycle (see Fig. 8.1).

There are two important techniques that have been employed in man to investigate the degree to which circadian systems are controlled and coupled to each other.

The first technique is to remove environmental or social cues regarding the planet's day–night cycle by seclusion in a specially constructed laboratory. During this seclusion, which if the experiment is to be useful must last several weeks, the biological rhythms are presumed to be "running free" of external factors. Monitoring physiological and psychological variables then provides information about the inherent properties of the oscillators that generate them.

Data from such studies by Weitzman's group[10] in New York and classical studies in caves and bunkers by Halberg and associates,[11,12] Kleitman,[13] Aschoff,[14] and others indicate that frequently the rest–activity cycle lengthens (i.e., it adopts a period greater than 24 hours). Further, the various parameters of the circadian cycle that normally run in phase with each other begin to split apart. While temperature usually stays close to a 24-hour period, the sleep–wake pattern may move to a 25- or 26-hour cycle, so that over a number of days and weeks the temperature cycle and the sleep–wakefulness cycle become increasingly out of phase. When this happens, a sudden change in activity frequently occurs, with the individual choosing to stay awake for a longer period than is their habit. Interestingly, however, when they do eventually settle down to sleep, it is as the body temperature begins to drop. A phase reintegration of the temperature and sleep–wake cycles is thus achieved. Wehr of the National Institute of Mental Health has observed that such "long days" are reminiscent of the behavior of individuals switching from mania to depression; indeed, preliminary investigation of individuals with rapidly cycling bipolar illness does indicate that the sleep–wake cycle and the circadian temperature rhythm may have a shifting phase relationship through the different periods of the illness.[15,16] The data from free-running experiments suggest that although most of the rhythms measured appear to have an endogenous circadian period, it is not completely autonomous of the environment. External markers or *zeitgebers* (literally "timegivers," a term first used by Aschoff) are still necessary to ensure precise synchronization of internal events with the environment; hence the rhythms are indeed *circa* (about) *dies* (24 hours) rather than being precisely so. The most powerful *zeitgeber* in plants and animals is the light–dark cycle. Although in human beings there are many other clues which probably help synchronize the clocks, there is increasing evidence that light remains of considerable importance in entraining our own rhythmic physiology.

The second type of experiment utilizes a rapid shift of the waking period within the 24-hour day as a perturbation of endogenous rhythmicity. Rapid travel across time zones or the reversal of the sleep–activity cycle by night work are the most obvious natural experiments in ourselves. Many people have experienced the subjective disorganization that such shifts induce, and learning more about the phenomena involved has been important for air crew performance, work productivity, and personnel safety. Some interesting points have emerged. For example, psychobiological adjustment to west–east flight is more difficult than the adjustment when flying from east to west. While the rest–activity cycle through social necessity may appear to be quickly synchronized (regardless of the direction flown), the subjective and physiological effects of "jet lag" are correlated with the direction of travel. Temperature and heart rate require about 4 days for reentrainment on east–west flight and 4 to 6 days on west–east flight. Mental performance seems to follow temperature, and a performance deficit is noted when the temperature cycle is at its low ebb. Physiological parameters such as palmar evaporative water loss take even longer (about 8 days) to reentrain, with the differences between east and west flight persisting.[17–21] The probable explanation of these differences lies in the tendency of the internal clocks to adopt a period that is longer than the 24-hour light and dark cycle of the sun. This preference, "to run a little slow," was noted in the free-running experiments outlined earlier. When traveling from the west to the east, personal time is "lost" and the demand to resynchronize comes even earlier than usual in the individual's day; when moving west, however, from the perspective of one's personal clock, time is "gained," accommodating the natural tendency of the internal oscillators to slow down and run with a period slightly greater than 24 hours.

The varying speed at which reentrainment occurs for different physiological parameters also indicates an hierarchy and interdependence of the internal rhythms and their sensitivity to environmental cues. This has particular importance in the work place, and the data indicate that accidents increase and mental efficiency decreases if demanding tasks are undertaken before reentrainment of the temperature rhythm with the rest–activity cycle has satisfactorily occurred.[22]

Light as a Zeitgeber in Human Beings

Just how important is light as a *zeitgeber* in human beings? After all, we have used artificial light for many years now to extend the active period of the day—with seemingly little physiological effect. Interestingly, recent information suggests that the intensity of the artificial light we use in our homes is insufficient to act as a synchronizer. Either sunlight or intense artificial light (of approximately 3000 lux and a similar spectral wave length to the sun) is required. Early experiments suggested that when an individual sleeps in the dark and remains blindfolded during the first 3 hours while up and about, there is a phase delay in the circadian rhythm of electrolyte and water secretion. More recently Lewy and associates have demonstrated in man that sunlight and bright artificial light will suppress pineal melatonin excretion.[23] In animals, it has been recognized for some time that this gland is part of an important mechanism for photic entrainment of behavior. In rats (and there is probably a similar system operating in man) light appears to exert its organizing influence partly through an effect upon the pineal gland's biogenic amine metabolism which in turn controls the secretion of melatonin. Wurtmann,[24] Axelrod,[25] and others have shown that this indoleamine is produced in the pineal by the acetylation and then methylation of serotonin (see Fig. 8.2). Light impinging upon the retina stimulates the transmission of photic information to the gland via a nonvisual pathway called the inferior accessory optic tract. This pathway crosses the optic chiasma and passes via the suprachiasmatic nucleus (where most of the endogenous clocklike functions of the CNS originate) to the medial forebrain bundle, eventually ending in the superior cervical ganglion. Here lie the cells that send fibers to the pineal. The tract appears to inhibit melatonin production during daylight when high levels of serotonin can be detected in the pineal. In darkness norepinephrine is released which via a beta-adrenergic receptor mechanism stimulates adenylate cyclase and the synthesis of the enzymes that manufacture melatonin. A precise circadian rhythm of melatonin synchronized with the light–dark cycle is the result. The most obvious metabolic effect of melatonin in animals is the inhibition of gonadal development and activity. What other effects it has on endocrine systems are unclear, but that it has a definite influence upon thyroid metabolism now seems likely. In man the daily pattern of melatonin excretion offers an excellent marker of the central nervous system's response to the light–dark cycle, and with methods of melatonin assay now improving, measurement of melatonin excretion in human subjects has become of considerable interest.

The Changing Photoperiod and Seasonal Depressions

While the most obvious motion of the earth is its daily rotation, determining the circadian light–dark cycle, the 600-million-mile journey around the sun also varies over a year's time the amount and intensity of light reaching any one point on the planet's surface. It is the 23.5° tilt of the earth relative to the plane of its orbit that determines this changing circa-annual (circannual) photoperiod; adaptive mechanisms to accommodate these changes are to be found throughout living things. In animal

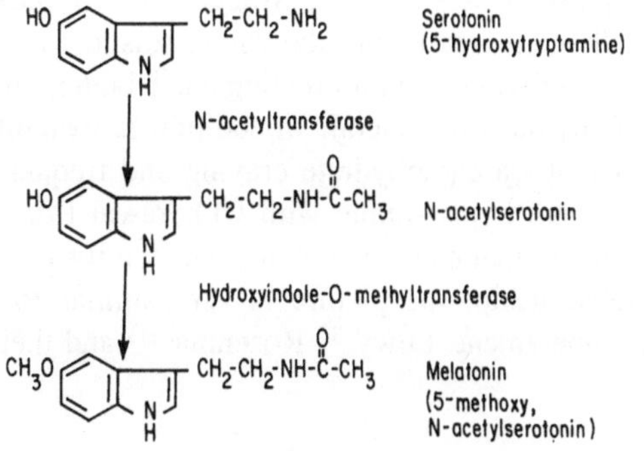

Figure 8.2 Synthesis of melatonin.

behavior, for example, sexual activity in those creatures with extended gestation periods is strongly influenced by the photoperiod. Herd animals such as sheep and deer are most sexually active in the fall when the photoperiod is declining, thus ensuring that offspring will be born predominantly in the spring.[26] Testosterone levels in the males and the growth of antlers in bucks of the deer herd are tied to this annual rhythm, which may be altered under experimental conditions by altering the photoperiod.[27] One report[28] suggests that a similar circannual rhythm in sexual activity may occur in human males living in northern Europe, and there is evidence that the number of summertime births increases with latitude.[29]

Obviously, where one lives on the earth's surface makes a big difference in the changes in the photoperiod that must be accommodated over this annual cycle. At the equator there is maximal change in amplitude of the variation in any 24-hour period, whereas at the poles that change is minimal. During any one day it is either dark virtually all the time, as in winter, or light all the time, as in summer. If one looks at the changes over the complete year, however, the reverse pertains. The amplitude of change is greatest at the poles and minimal at the equator. The earth's climate therefore becomes more "seasonal" as one moves toward the polar regions. Logically, it also follows that annual variation in endogenous rhythms entrained to the changing light patterns will be influenced by the latitude at which one lives.

Seasonal depressions, a phenomenon noted in earlier chapters, have been recognized for a long time. "Of the seasons of the year the autumn is most melancholy," wrote Robert Burton in 1621.[30] We also noted the peaks of successful suicide in New England to be in the spring and the fall (Fig. 1.2). Interestingly, similar patterns are found in the southern hemisphere which correspond with the comparable seasons, and in both hemispheres the periodic incidence of the phenomena becomes more pronounced with latitude.

Could it be that these seasonal variations in psychopathology are tied to the larger environmental change, perhaps specifically to changes in the photoperiod? The assay of melatonin as a reliable index of photic influence upon the central nervous system has made it possible to explore this question. Dawn light rapidly suppresses melatonin secretion in healthy individuals, as will artificial light of sufficient intensity (between 2000 to 3000 lux). There is some evidence that in manic–depressive illness melatonin is more easily suppressed, suggesting that the rhythm is less well coupled to the endogenous oscillator than in normals.[31]

Individuals with seasonal depressions are less severely ill than many depressives but share features in common with those in the depressed phase of bipolar illness. Approximately 75% are women. In the typical case the symptoms usually begin in the fall and increase until the winter solstice, following which a slow remission develops. The individuals report feeling much better during the summer months and during bright sunny days in general. In addition to irritability, sadness, and some retardation, they describe a carbohydrate craving and frequently gain weight during the winter. They also are hypersomnic with withdrawal from elective social activities, but most individuals maintain their work and family responsibilities. Save for the clear seasonal relationship, the picture is very similar to that of an individual with cyclothymic temperament. Lewy,[32] Rosenthal,[33] and their co-workers recently have reported preliminary studies in which such individuals were exposed to either 3000 lux of simulated sunlight morning and evening or to a comparable period of yellow light. Self-ratings of mood suggested improvement in mood during the exposure to bright light, tending to confirm the personal experience of many of these patients who have found migration south in the winter to be an effective antidepressant.

These results, although fascinating, are very preliminary, and thus cautious interpretation is in order. Also, those with seasonal depressions are a very small percentage of the total population of depressed persons. Nonetheless, the findings suggest that it may be useful to structure affective illness conceptually as a disorder of central nervous system regulation. If, as we suggested earlier, one accepts that the main function of the central nervous system is maintaining spatiotemporal order, then the orchestration of interdependent rhythms is a vital function in determining psychobiological health. This function would be enhanced by the precise entrainment of the rhythms to the light–dark cycle, which could then act as a daily synchronizer. In winter, in the absence of an adequate photic stimulus, some individuals in whom the rhythms were poorly coupled would be more prone to physiological disorganization and thus depression. Kripke has suggested that there is a critical interval for photic stimuli to sustain entrainment in the early morning and that this period may lie outside the natural photoperiod in winter[34] (Fig. 8.3). An alternative explanation is perhaps that social necessity forces us to rise before dawn in winter and the entrainment induced by the dawn (marked by the fall in secretion of pineal melatonin) does not occur with precision. In either instance, the absence or blunting of the photic stimulus to entrainment allows further uncoupling of biological oscillators. In some individuals this is experienced as seasonal depression and exhibited physiologically as loss of the usual phase relationships between endogenous rhythms.

Common experience gives some support to this conjecture. When artificial light is available, bedtime for most of us is usually long after sunset, but wake-up time remains around dawn or shortly after. In contrast, there are very few who by choice rise during the dark hours of the early morning and go to bed before the sun sets. Together with the preliminary experimental evidence, this suggests that photic entrainment remains important even in human beings. It is an exciting area of research which, if found to be

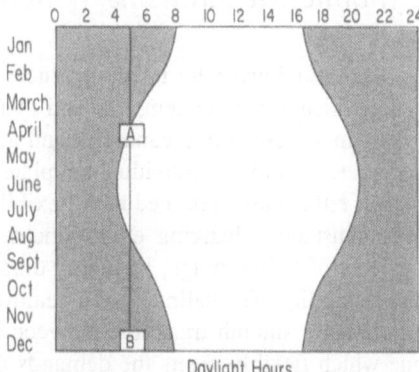

Figure 8.3 Seasonal variation in the photoperiod. Living at a latitude of 40° and above there is a considerable variation of the photoperiod during the annual cycle. Kripke has suggested that there is a crucial period for entrainment of biological rhythms by light; during the spring and summer (A) this period occurs around dawn, but in the middle of winter dawn occurs after the critical period has passed (B).[34]

valid, will have important implications for social behavior and the work schedules that we adopt.

Evidence for Temporal Disorder in the Final Common Path of Affective Illness

We have reviewed circadian rhythmicity and seasonal psychopathology at some length to underscore the potential insights that may develop when exploring disorders of mood from a dynamic standpoint. It is possible that mood disorders, representing a final common path of psychobiological dysfunction, are syndromes in which temporal regulation has been lost; during illness the stable dynamic organization that usually facilitates environmental interaction is temporarily disturbed and flexibility and adaptation are consequently impaired. There is considerable evidence that the usual regulation of psychophysiological events may be lost in pathological affective states.

The usual internal architecture of the circadian cycles is impaired in depression, for example, and also to some extent in mania.[35,36] There is a shift in the phase relationships between rhythms. The peak of 17-hydroxycorticosteroid (17-OHCS) excretion occurs earlier and is disassociated from the temperature cycle. Sleep architecture is radically altered with a decreased rapid eye movement latency, a reduction in slow-wave sleep, and total sleep time. In symptomatology there is frequently a marked diurnal variation with the syndrome being worse in the early morning. In those individuals with rapidly cycling illness, the switch process from depression to mania also frequently occurs during the early morning hours. All these phenomena appear to indicate a loss of the usual dynamic interplay and temporal relationships among key physiological variables in affective illness.

Dynamic Mechanisms Which Serve Stability and Flexibility

An individual who is coping successfully has a range of psychobiological templates which permit the temporal and spatial organization of internal and environmental events in a personally meaningful and adaptive way. This organization is not perfect, however, nor are the individual templates or models locked into an immutable oscillation. If either case pertained then flexibility would be lost to a stability that in the face of a constantly changing environment would be meaningless. A most remarkable dynamic of living things, including ourselves, is the ability to maintain both stability *and* flexibility. The hallmark of a healthy person is the capacity to orchestrate through regulation a smooth transition between a preferred stable psychobiological state and one which flexibly meets the demands of a new perturbing environment.

The combination of regulatory mechanisms that underly the ability to adapt to a seasonal environment is an excellent illustration of this important point. The reader will recall that the endogenous sleep–wake cycle in a free-running individual characteristically has a period slightly *longer* than 24 hours. Final adjustment of the period to ensure the best possible fit with environmental circumstance is achieved by reentrainment each morning to the photic stimulation of the rising sun. Stability and flexibility are assured by the combination of an internal template (in this case a clocklike mechanism probably located in the suprachiasmatic nucleus) providing basic organization and a daily cue (photic stimulation) which provides information regarding continuously changing external circumstances. If the internal oscillator were more easily manipulated by environmental signals or were so fixed in its cycles as to be impervious to external circumstance, then flexibility would be lost and a predisposition to pathology would emerge as smooth adaptation to the environment became impossible.

In those prone to develop affective illness, for example, a predisposition of endogenous oscillators to drift may be reflected in the greater ease with which melatonin is suppressed by light in bipolar persons or the persistence of abnormalities in sleep architecture even after recovery from illness. A locked periodicity as seen in rapidly cycling manic–depressive illness represents, on the other hand, a locked oscillator which is very resistant to perturbation by the social environment or pharmacotherapy. In either instance, the usual phase relationships among rhythmic physiological functions have been lost and the temporal organization of internal and external events becomes seriously impaired.

Dynamic Systems and Subsystems: Some Definitions

These conceptual considerations regarding circadian and seasonal rhythms offer a general paradigm for integration of the other information we have about affective illness. Key to that integration is recognition that mood disorders are diseases of the total system; the events around the synapse are disordered just as are interpersonal relationships. Furthermore, the set of forces that pertains to one subsystem may in fact sustain the disorganization and pathology present in another. The parts interact to create the whole, and the whole individual in turn is in dynamic interaction with the environment. A difficult infant–parent relationship and subsequent separation can therefore theoretically equally well set the stage for adult depression as can a genetically based disorder of neurotransmitter dynamics at the synapse. Each element must be considered part of the whole.

In thinking about complex dynamic (that is spatiotemporal) relationships, it is important to recognize that a dynamic system is only what we choose to define it to be. The dynamics of biogenic amine metabolism as they pertain to the transport of information across the synapse, the economics of a family, the interplay of mechanisms of psychobiological defense in response to the loss of an important person—all may be defined and investigated as complex systems. While each is part of the whole, the dynamics can be explored separately. Arthur Winfree, in his book *The Geometry of Biological Time*, has put it succinctly:

> Although it may be fashionable to acknowledge that everything is connected to everything else in principle, some things are more tightly connected to each other than all the rest. Such a little knot of causal interactions goes by the name of a system . . . the

state of a system consists of everything you need to know about it right now in order to know what it will do next in response to select stimuli . . . . Dynamical systems have the interesting property of continually, spontaneously changing state. [At any time point] the system's state is determined by a collection of variables, things which [impinge upon the system and] change spontaneously. Things [which help define the system but] don't change spontaneously are called parameters.[8] (p. 74)

The overall behavior of any system reflects the advantages for or constraints upon flexibility imposed by its parameters (e.g., genetic programming and acquired psycho-biological characteristics, in the case of the individual with mood disorder) and the variable demands of the environment external to the system itself. Exactly where the line in any system is drawn between system and environment is a matter of conceptual preference (see Fig. 8.4). Over many years our familiar divisions in the life sciences such as biochemistry, physiology, psychology, and sociology have developed around methods of observation, and common experience as dictated by prevailing philosophy and available technology. The result has been a pragmatic clustering of phenomena and at times a spurious division of naturally integrated functions. Our theoretical models, the orderly description of selected subsystems, have followed these established patterns, and appear superficially to be grossly divergent and competitive in their explanations of the phenomena observed. The several models of mood disorders which we outlined in Chapter 2 are an example of this. However, in fact, they represent segments of the same dynamic whole, individual subsystems that are interdependent. As each *is* part of the whole, the behavior described in each of these conceptual subsystems should be dynamically similar and lawful, with a common logic and geometry.

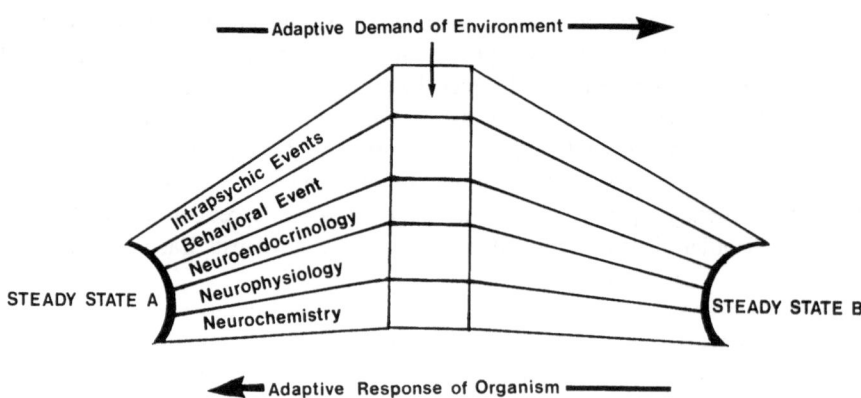

Figure 8.4 All living organisms through the consumption of energy maintain a steady state (in our diagram: position A) and resist perturbation to a new steady state (i.e., pathological state) such as is represented in our diagram by position B. As we view the constant flux of the organism (or individual) we choose to divide it for study into various elements, e.g., behavioral, neuroendocrine, etc. However, none of these is truly distinct. They are merely a scientific convenience.

Describing the Behavior of Complex Systems: Mathematics as a Potential Common Language

It follows from what we have outlined that when describing a dynamic system disorder such as affective illness, it is not only important to investigate the parts but also the functional relationship between those parts. In describing the disorder, therefore, it is important to identify the underlying forces, and the temporal relationship they have to each other as the behavioral disturbance is developed and sustained. How does one achieve this for so many diverse sets of information?

In most instances the real origins of the forces and the structural mechanisms that underlie the behavior of living creatures are unknown. However, as we have seen, the behavior itself is lawful. Monitored over time, for example, recurring patterns become apparent, the properties of which can be described mathematically. Hence the hope has frequently been expressed that mathematics may offer the language necessary to describe in common idiom the complexities of psychobiology and psychopathology.

In recent years there have been several attempts to develop a mathematics of the general behavior of complex systems. The general system theory of Von Bertalanffy was one of the first and remains popular.[37] Dynamics, a discipline that lies between mathematics and the sciences, offers a similar opportunity for abstraction through the development of a geometry of behavior.[38]

A common language has great appeal. It would certainly also have great utility, potentially transcending our historical and essentially arbitrary classifications of the phenomena that we observe in existing basic science and clinical disciplines. It would be possible for investigators of differing technical expertise and theoretical persuasion to find common conceptual ground in the dynamics of the behavior exhibited by the system they had under study. When viewed as a functional unit, the concept of ego and its mechanisms of defense may indeed be capable of describing behavior in a manner lawful and similar to those fundamental rules that govern the dynamic flow of events around the neuronal synapse! Mathematics would become the ultimate integrator!

Sadly, such Utopian visions are still some way off. Certainly computer technology has now made it possible to visualize the temporal morphology of complex systems by computer graphics. Geometric theories of behavior in space and time can be developed, tested, and accurately represented as never before. Because the behavior of complex systems is indeed so complex and well beyond the imagination of the average individual, such technical aids do have some utility and can provide helpful new insights into well-established but unexplained phenomena.

Let us take as an example "switching" from one state to the other in manic–depressive illness. Any model that seeks to explain the behavior commonly found in bipolar disorder must be able to accommodate the switch process in which suddenly mania appears like a phoenix from the ashes of depressive disturbance. Observation suggests that continuously changing, apparently random events precede a major discontinuity in behavior. In other instances, the switch appears to be induced by one of a broad range of factors such as antidepressant therapies or by endocrinological manipulation, principally thyroid hormone or 17-OHCS treatments.

Seeking to better understand the phenomenon through the perspective of dynamic

theory, we first adopt a working view of brain as a collection of hierarchically organized and strongly coupled oscillators which continuously drive one another. They
each have a preferred steady state or stability but are capable of rapid change in
response to environmental challenge. Variables impinge constantly upon the system (in
this case a manic–depressive person), perturbing these oscillators; despite inherent
stability, one oscillatory subsystem may be driven by repeated perturbation to the point
where it entrains those subsystems around it and forces a dramatic change in the
functions of all of them together. A sudden universal change in the behavior of the
whole system then occurs. To the observer the system's behavior changes suddenly
and in a "catastrophic" way. A mathematical method which can lawfully describe
such catastrophic behavior was created by Rene Thom is 1972 and made popular by
Zeeman.[39,40] It is a general dynamic theory which explains how gradually changing
forces can produce discontinuities, or sudden effects.

Thus we learn from this theory that the switch process does not need to be viewed
as a unique phenomenon—rather it is one predictable dimension of the behavior of
complex systems composed of driven coupled oscillators. Translating back into biological language, any number of variables impinging upon the system could induce mania
if the phase relationships between rhythmic functions such as those of the catecholamines, receptor sensitivity, and endocrine mechanisms were conducive to the
change. So, for example, thyroid hormone, although initially helpful in the depressed
individual, may by induction of adrenergic receptor activity lower the critical threshold
for mania.[41] The peak of the daily oscillation in available synaptic catecholamines
would then be sufficient to trigger the chain of events leading to the development of the
new, biologically based steady state of manic behavior. The system literally "steps"
from one preferred behavior and adopts another.

Conclusion and a Review of Dynamic Constructs

Whether the mathematical analysis of the behavior of dynamic systems will
become of practical value to research scientists and clinicians remains to be seen.[42]
Exceptional collaboration between mathematicians and biologists will be required
before a satisfactory level of mutual understanding is reached. It is not that any
biologist or clinician doubts the complexity of the system or disorder under study;
Claude Bernard convinced every interested person of that 100 years ago. No, the
central problem remains today what it was then. To investigate the detailed mechanisms that underpin a complex living organism, one must inevitably conduct research
upon its parts; but how does that information relate to the behavior of the whole? The
development of an abstract common language solves only part of the problem, and we
cannot attribute the burden of our lack of understanding to its absence.

Because living creatures are the sum of their interactions, however, a dynamic
perspective, although basically alien to us as a way of thinking (remember how we
prefer to separate mind from body!), can provide some useful cognitive guidelines.
This is regardless of whether the concepts are struggled over in the abstract language of
mathematics or in prose.

What follows is a review, in note form, of some of the dynamic constructs that pertain to the behavior of complex systems and that we have found useful in our own thinking. Some have been discussed in this chapter, and some have been utilized and referred to in our earlier review chapters. The concepts surrounding the regulated response of complex systems to perturbation and the development of a final common path of disorder will have particular pertinence to our next chapter, in which we seek to integrate current information regarding affective illness.

1. *A dynamic system* is a set of causal interactions.
   - The limits of a system are as we choose to define them.
   - Within one complex system many subsystems may be defined.
   - Conceptually systems cross descriptive barriers such as that of mind and body.
   - The parameters of a system are those relatively stable elements, knowledge of which helps predict the system's future interaction (behavior).
   - The variables of a system are those elements that change spontaneously and in doing so influence its interactional behavior.
2. *Living organisms* are open systems.
   - They exhibit a high degree of autonomy while in constant commerce with their environment.
   - Their survival is keyed to the ordering of internal and external events in time and space.
   - Thus regulatory mechanisms are essential to the preservation of all living organisms; these mechanisms control the flow of information between functional units (cells or subsystems) and orchestrate the parts of the whole.
   - A centralized nervous system is the ultimate regulatory mechanism; it serves homeostasis through feedback of information upon itself and development of spatiotemporal models (templates) of the external environment.
   - It has been suggested that a mechanical analogy for understanding the behavior of the central nervous system is that of a computer with well-developed clocklike mechanisms.
3. *The behavior of a living organism* is nonlinear but lawful.
   - The information gathered from the interaction among subsystems and between the organism and the larger environment is constantly reentered upon itself. A collection of inherently stable continuous cycles of behavior is the result.
   - These cycles of behavior when monitored over time appear as rhythms. For conceptual and analytic convenience, each rhythmic cycle may be seen as the product of a nonlinear oscillator.
   - These oscillatory (dynamic) behaviors are open to mathematical analysis.
   - Oscillating subsystems are frequently coupled with each other (i.e., they are highly interdependent). They also have the property of becoming entrained to other oscillators, such as those that may dominate in the external environment (e.g., the circadian or seasonal cycles dictated by the earth's changing relationship to the sun).
   - The subsystems are hierarchically organized with considerable overlap and redundancy. The manifest behavior of an organism flows from the balance of these many subsystems which frequently function in opposition to each other.

Such opposition affords a sensitive mechanism for fine tuning the living organism's interaction with the environment.

4. *Interaction between an organism and its environment* is essential to survival.
   - From this interaction spatiotemporal models (templates) of the environment are developed; they are interactional analogues of the real world.
   - Such templates are an important form of planning: they help maintain homeostasis and conceptually they form an integral part of all biological and psychosocial systems.
   - Some have been developed over evolutionary time and incorporated into the genetic endowment (e.g., the clocklike mechanisms of the suprachiasmatic nucleus). Others develop on individual need and in response to specific environmental demand (e.g., interactional patterns identified as characteristic of nurturing, immunologic identification of bacteria, concepts of ego ideal).
   - In either instance the models have flexibility at their margin (e.g., entrainment of circadian rhythms by photic information, modifications of ego ideal dependent upon experience and personal performance). Such continued modification of the basic template permits ongoing adaptation to a changing environment.
   - These templates collectively determine a set of personally preferred psychobiological norms. These contribute to the parameters within which the system is capable of interaction without major burden.
   - The templates are dynamic in that, together with all other aspects of the dynamic system, they are dependent upon the quantity and quality of interaction the individual has with the environment. With little interaction the template deteriorates through lack of reinforcement; continued interaction facilitates continued refinement and development.

5. *Perturbing variables impinge constantly* upon a living organism.
   - Information pertinent to these variables is continuously appraised against an individual's accepted norms in space and time (templates).
   - The variable may be considered deviant or novel if the information is unfamiliar (i.e., differs in quality from that previously experienced) or if the quantity of information received in unit time is radically different from the temporal norm (e.g., in the mother–child interaction the quality of the information changes if the mother is replaced by another caring person, and the quantity of interactional information drops precipitously if the mother leaves and does not return).
   - Deviation promotes adaptational change, of which there are two basic possibilities: (1) modifying the perturbing variable (e.g., protest in the case of infant separation, which is designed to return the mother to the interaction) and (2) modification of the template norm (pursuing the example of the mother–child interaction, if the mother does not return the child withdraws, modifying psychobiological goals and associated norms; such behavior was exhibited by Monica, see Chapter 4).
   - Adaptational changes are effected by the basic psychobiological strategies (coping behaviors) available to the individual. Again, these are based upon genetic and acquired skills with emergency augmentation at their margin depending upon the nature of the demand.

- If the adaptational response is successful, homeostasis is maintained, and template norms and coping strategies are modified on the basis of the experience.

6. *If coping strategies are inadequate and the perturbing influence continues:*

- Over time a final common path of dysfunction begins to emerge.
- The system loses its flexibility and its ability to interact with the environment; individual oscillators begin to uncouple or couple in new ways (e.g., a dissociation between the sleep–wake cycle and the temperature cycle, as seen in depression).
- The overall behavior of the system changes, sometimes suddenly; the old norms of behavior are lost or considerably modified within the parameters of the new steady state (e.g., melancholia).
- The phase relationships of the dynamic functions characteristic of health are lost; an apparent disorganization replaces the usual orchestration of psychobiological function, or a locked periodicity occurs which is impervious to environmental perturbation (e.g., rapid manic–depressive cycling).
- The new steady state may persist for weeks or months, and over that time the diminished interaction with the environment may result in further disorganization or destruction of the individual.

References

1. Tolstoy, L. *The death of Ivan Ilych, and other stories.* New York: New American Library of World Literature, 1960.
2. Popper, K. R., and Eccles, J. C. *The self and its brain.* New York: Springer International, 1977.
3. Bernard, C. English edition translated by H. C. Greene. *An introduction to the study of experimental medicine.* New York: Macmillan, 1927.
4. Haldane, E. S. *Descartes, his life and times.* London: J. Murray, 1905.
5. Young, J. Z. *A model of the brain.* London: Oxford University Press, 1964.
6. Oakley, K., and Goodwin, B. C. The explanation and investigation of biological rhythms. In W. P. Colquhoun (Ed.), *Biological rhythms and human performance.* London: Academic Press, 1971, pp. 1–38.
7. Pittendrigh, C. S. Circadian oscillations in cells and circadian oscillation of multicellular systems. In F. O. Schmitt and F. G. Worden (Eds.), *The neurosciences,* Third Study Program. Cambridge, Massachusetts: MIT Press, 1974.
8. Winfree, A. T. *The geometry of biological time.* New York: Springer-Verlag, 1980.
9. Halberg, F. Implications of biological rhythms for clinical practice. In D. T. Krieger and J. C. Hughes (Eds.), *Neuroendocrinology.* Sunderland, Massachusetts: Sinauer, 1980, pp. 109–119.
10. Czeisler, C. A., Weitzman, E. D., Moore-Ede, M. C., Zimmerman, J. C., and Knauer, R. S. Human sleep: Its duration and organization depend on its circacian phase. *Science,* 1980, *210,* 1264–1267.
11. Halberg, F., Reinberg, A., Haus, E., Ghata, J., and Siffre, M. Human biological rhythms during and after several months of isolation underground in natural caves. *National Speleology Society Bulletin,* 1970, *32,* 89–115.
12. Halberg, F. Biological rhythms. *Advances in Experimental Medicine and Biology,* 1975, *54,* 1–41.
13. Kleitman, N. *Sleep and wakefulness,* 2nd ed. Chicago: University of Chicago Press, 1963.
14. Aschoff, J. Exogenous and endogenous components in circadian rhythms. *Cold Spring Harbor symposia on quantitative biology,* 1960, *25,* 11–28.
15. Wehr, T. A., and Wirz-Justice, A. Circadian rhythm mechanisms in affective illness and in antidepressant drug action. *Pharmacopsychiatria,* 1982, *15,* 30–38.
16. Wehr, T. A., and Goodwin, F. K. Biological rhythms and psychiatry. In S. Arieti and H. K. H. Brodie

(Eds.), *American handbook of psychiatry,* vol. 7, 2nd ed. New York: Basic Books, 1981, pp. 46–74

17. Klein, K. E., Wegmann, H. M., Athanassenas, G., Hohlweck, H., and Kuklinski, P. Air operations and circadian performance rhythms. *Aviation, Space and Environmental Medicine,* 1976, *47*, 221–230.
18. Rutenfranz, J., and Knauth, P. Investigation of the problems concerning influences upon the sleep of shiftworkers. In A. Swennson (Ed.), *Night and shiftwork, Studia Laboris et Salutis,* 1969, *4*, 58–65.
19. Hauty, G. T., and Adams, T. Phase shifts of the human circadian system and performance deficit during the periods of transition. I. East–west flight. *Aerospace Medicine,* 1966, *37*, 668–674.
20. Hauty, G. T., and Adams, T. Phase shifts of the human circadian system and performance deficit during the periods of transition. II. West–east flight. *Aerospace Medicine,* 1966, *37*, 1027–1033.
21. Athanassenas, G., and Wolters, C. L. Sleep after transmeridian flights. In A. Reinberg, N. Vieux, and P. Andlauer (Eds.), *Night and shift work: Biological and social aspects.* New York: Pergamon, 1981, pp. 139–147.
22. Colquhoun, W. P. Circadian variations in mental efficiency. In W. P. Colquhoun (Ed.), *Biological rhythms and human performance,* London: Academic Press, 1971, pp. 39–107.
23. Lewy, A. J., Wehr, T. A., Goodwin, F. K., Newsome, D. A., and Markey, S. P. Light suppresses melatonin secretion in humans. *Science,* 1980, *210,* 1267–1269.
24. Wurtmann, R. L. The pineal as a neuroendocrine transducer. In D. T. Krieger, and J. C. Hughes (Eds.), *Neuroendocrinology.* Sunderland, Massachusetts: Sinauer, 1980, pp. 102–108.
25. Axelrod, J. The pineal gland: A neurochemical transducer. *Science,* 1974, *184,* 1341–1348.
26. Lincoln, G. A., Peet, M. J., and Cunningham, R. A. Seasonal and circadian changes in the episodic release of follicle-stimulating hormone, luteinizing hormone and testosterone in rams exposed to artificial photoperiods. *Journal of Endocrinology,* 1977, *72,* 337–349.
27. Goss, R. J., Dinsmore, C. H., Grimes, L. N., and Rosen, J. K. Expression and suppression of the circannual antler growth cycle in deer. In E. T. Pengelley (Ed.), *Circannual clocks.* New York: Academic Press, 1974, pp. 393–422.
28. Lagoguey, M., and Reinberg, A. Circadian rhythms in plasma LH, FSH and testosterone and in the sexual activity of healthy young Parisian males. Proceedings of the Physiological Society, 9–10 January, 1976. *Journal of Physiology,* 1976, *257,* 19–20.
29. Cowgill, U. M. Season of birth in man: Contemporary situation with special reference to Europe and the Southern Hemisphere. *Ecology,* 1966, *47,* 614–623.
30. Burton, R. (1621). *The anatomy of melancholy.* New York: Vintage Books, 1977.
31. Lewy, A. J., Wehr, T. A., Goodwin, F. K., Newsome, D. A., and Rosenthal, N. E. Manic depressive patients' may be supersensitive to light. *Lancet,* 1981, *1,* 383–384.
32. Lewy, A. J., Kern, H. A., Rosenthal, N. E., and Wehr, T. A. Bright artificial light treatment of a manic–depressive patient with a seasonal mood cycle. *American Journal of Psychiatry,* 1982, *139,* 1496–1498.
33. Rosenthal, N. E., Sack, D. A., Gillin, J. C., Lewy, A. J., Goodwin, F. K., Davenport, Y., Mueller, P. S., Newsome, D. A., and Wehr, T. A. Seasonal affective disorder: A description of the syndrome and preliminary findings with light treatment. *Archives of General Psychiatry,* 1984, *41,* 72–80.
34. Kripke, D. F. Photoperiodic mechanisms for depression and its treatment. In C. Perris, G. Struwe, and B. Jansson (Eds.), *Biological psychiatry 1981.* Amsterdam: Elsevier, 1981.
35. Kripke, D. F., Mullaney, D. J., Atkinson, M., and Wolf, S. Circadian rhythm disorders in manic–depressives. *Biological Psychiatry,* 1978, *13,* 335–351.
36. Wehr, T. A. Circadian rhythm disturbances in depression and mania. In F. M. Brown, and R. C. Graeber (Eds.), *Rhythmic aspects of behavior.* Hillsdale, New Jersey: Lawrence Erlbaum, 1982.
37. Von Bertalanffy, L. *General system theory.* New York: George Braziller, 1968.
38. Abraham, R. H., and Shaw, C. D. *Dynamics, the geometry of behavior.* Santa Cruz, California: Aerial Press, 1982.
39. Thom, R. Structural stability, catastrophe theory and applied mathematics. *Society for Industrial and Applied Mathematics Review,* 1977, *19,* 189–201.
40. Zeeman, E. C. Bifurcation, catastrophe, and turbulence. In P. S. Hilton and G. S. Young (Eds.), *New directions in applied mathematics.* New York: Springer-Verlag, 1982, pp. 109–153.
41. Whybrow, P. C., and Prange, A. J. A hypothesis of thyroid-catecholamine-receptor interaction: Its relevance to affective illness. *Archives of General Psychiatry,* 1981, *38,* 106–113.
42. Kac, M. Dehydrated elephants revisited. *American Scientist,* 1982, *70,* 633–634.

9

Toward a Psychobiological Integration: Affective Illness as a Final Common Path to Adaptive Failure

Physiologists and physicians must always consider organisms as a whole and in detail at one and the same time without ever losing sight of the peculiar conditions of all the special phenomena whose resultant is the individual. It is generally agreed that synthesis reunites what analysis has divided and that synthesis, therefore, verifies analysis of which it is merely the counterproof or necessary complement.

Claude Bernard[1]
An Introduction to the Study of Experimental Medicine (1865)

Introduction

In the late 1930s the philanthropist William T. Grant provided the financial means to study a group of male university students through the normal life cycle.[2] Many of these men are now in their early 60s. Initially individuals were included in the study because they appeared healthy, and although all have experienced challenge and disappointment, many have stayed so. Reviewing something of what has been learned through study of the lives of these men, George Vaillant has observed that the most striking feature has been their ability to adapt to changing circumstance. Psychological health, Vaillant concludes, is realistic adaptation, orchestrating life's demands so that they fall within one's ability to manage. Adapting with some degree of success, or at least struggling on reasonably equal terms with one's environment, may be termed coping. Consistent coping brings mastery.

Depressed persons lose these skills. In their place stands a peculiar alteration in self-perception. The usual hope and struggle to control the environment has been given up. Negative thoughts impair the motivation to obtain reward from friendship. Seeking reinforcement from others to confirm that one's actions are useful or worthy no longer seems important. Self-esteem is very low. Whatever the nature of the stressful events that culminate in such states of despair, melancholia is accompanied by a temporary disruption of interpersonal, vocational, and ideological bonds. The individual's relationship with other human beings, with human institutions, and with the world has been diminished, and yet no change is sought. As the withdrawal progresses this maladaptive nonrelatedness becomes self-perpetuating. Interpersonal interaction declines further, and a significant loss of behavioral reinforcement occurs. The individual describes a pervasive sense of helplessness, hopelessness, and a preference for death.

It is our thesis that this syndrome, touching 10–15% of the population at some point in life, is the behavioral manifestation of a new psychobiological state, the outcome of a final common pathway[3] of various interlocking processes at the neurophysiological, biochemical, experiential, and behavioral levels. We suggest that the biological disturbance concomitant with the behavioral changes, in the language of neurophysiology, exists as a reversible regulatory impairment, both interoceptive and exteroceptive, of the limbic–diencephalic brain centers serving psychomotor activity, mood, reward, and arousal.[4,5] In this new state the individual loses many of the characteristics of adaptive health and becomes locked into a cycle of behavior which shows remarkable autonomy, self-reinforcement, pathological periodicity, and resistance to perturbation. In developing this integrative framework we have sought to accommodate the insights gained from the various perspectives that were reviewed in earlier chapters. The approach we have taken is an interactive one, considering the individual, both biologically and intrapsychically together with the larger social environment, as a general system. Each conceptual level of organization is considered linked and covariant with every other. Thus, while none of the models that were outlined in Chapter 2 are given primacy with regard to etiology of the syndrome, some of the elements of those models are found as components in this discussion.

Although we see melancholia as often precipitated by stressors which induce a disorder of central nervous system regulation, this is not equivalent to suggesting that the disability is environmentally determined. Environmental challenge contributes to the precipitation of many diseases, including depression, but conversely, as noted in the Grant study, not all those who are challenged become ill. Despite such daily perturbation, 85% of the population *does not* experience melancholia during a lifetime. The task is to catch the essence of that which is special about those who do.

Transient episodes of depressed mood are ubiquitous. The "permission" for serious disturbance, however, builds over many years and is contributed to by a multiplicity of factors, the combination of which varies among individuals. In some a genetic predisposition is amplified and shaped by acquired developmental parameters such as childhood separation and loss, physical illness, aging, and the interactional demand required by these various life challenges. In others in whom mania occurs as part of the syndrome, there may be an additional genetic component which determines the special course of the affective disturbance once it has emerged. During the developmental period and on into adult life, the adaptive capacity and style of the individual become increasingly constrained by these genetic and acquired patterns until a precipitating event or series of events disturbs the psychobiological equilibrium sufficiently to trigger the final common path to disorder.

As will be recognized from our earlier discussion of dynamic systems, interactional concepts of behavior do not lend themselves easily to linear prose. Thus to better order the description of this integrated theory of affective illness, we have chosen four main foci:

1. *The evidence for a final common path to disorder:* The factors that contribute to melancholia or mania vary among individuals, but the final constellation of psychobiological dysfunction is very similar despite this varied clustering. This suggests that the syndrome of melancholia (and mania) represents a final pathway—a common outcome—of those diverse predisposing variables. In this section we review its diversity from the clinical perspective and briefly summarize the manifest elements of the pathway itself.

2. *The predisposition to disorder:* Western scientific tradition fosters the assumption that predisposing factors are either psychological, social, or biological rather than a function that grows out of the interaction of all three influences. In reality, for each individual it is usually the mixing and working together of these elements that constitutes the predisposing variance. Separately they have little power, but together they create a vulnerability which is heterogeneous and dynamic. In this section we review the following interactive parameters:
   a. Genetic vulnerability
   b. Predisposing temperament
   c. The modulating influence of age and sex
   d. Physical illness and its treatment
   e. Loss of attachment in childhood
   f. Predisposing character traits

3. *Precipitating factors which initiate the path to dysfunction:* Predisposing parameters alone, by their very definition, are insufficient to cause illness; precipitating factors are required. Conceptually these factors perturb the dynamic function of the individual and threaten homeostasis. The time and nature of the challenge interact with individual vulnerability, and if adaptation is inadequate the path to disorder is initiated. Again biological and psychosocial variables may serve as precipitants, but in the human being psychological interpretation of the meaning of the event often plays a major role in determining the degree of perturbation that results. These concepts, and evidence regarding the nature of the psychosocial precipitants in affective illness, we have reviewed in earlier chapters. In this section we discuss some of those data in the context of the interactive model.

4. *Intermediary mechanisms:* In this final section the mechanisms thought to underpin the interaction of an individual with the environment are reviewed, with particular emphasis upon concepts generated by stress research. Central to the discussion is the thesis that the interaction of predisposing parameters and precipitating stressors initiates a disturbance of the regulatory mechanisms of the limbic area and the diencephalon at large.

The Evidence for a Final Common Path to Disorder

To the clinical observer the factors associated with the development of affective illness are varied. Depressed mood is frequently seen as a self-limiting adaptive response to *loss of attachment*, especially when it occurs in the context of romantic disappointment, geographic moves, or bereavement. *Drugs* commonly used in clinical practice, such as reserpine, alcohol, steroids, and the antipsychotic phenothiazines, are recognized as sometimes precipitating clinical depression.

A variety of *physical illnesses*,[6,7] including, for example, endocrinopathies such as hypothyroidism, hyperparathyroidism, Cushing's and Addisonian syndromes, occult abdominal malignancies, dementing neurological disorders, and collagen diseases such as systemic lupus erythematosus and rheumatoid arthritis, are also associated with the syndrome. Most *chronic debilitating illness*—whether physical or psychiatric—that causes pain and imposes limitation upon social interaction and the other modes of

Table 9.1 Somatic Disturbances and Drugs Associated with Depressive and Manic States

	Depression[6,7]	Mania[7,8]
Infectious	Influenza	Influenza
	Viral hepatitis	St. Louis encephalitis
	Infectious mononucleosis	Q fever
	General paresis (tertiary syphilis)	General paresis (tertiary syphilis)
	Tuberculosis	
Endocrine	Myxedema	Hyperthyroidism ?
	Cushing's disease	
	Addison's disease	
Neoplastic	Occult abdominal malignancies (e.g., carcinoma of head of pancreas)	
Collagen	Systemic lupus erythematosus	Systemic lupus erythematosus
		Rheumatic chorea
Neurological	Multiple sclerosis	Multiple sclerosis
	Cerebral tumors	Diencephalic and third ventricular tumors
	Sleep apnea	
	Dementia	
	Parkinson's disease	
	Nondominant temporal lobe lesions	
Nutritional	Pellagra	
	Pernicious anemia	
Drugs	Steroidal contraceptives	Steroids
	Reserpine	Levodopa
	Alpha-methyl-dopa	Amphetamines
	Physostigmine	Methylphenidate
	Alcohol	Cocaine
	Sedative-hypnotics	Monoamine oxidase inhibitors
	Amphetamine withdrawal	Thyroid hormones

interpersonal reinforcement is frequently accompanied by dysphoria, demoralization, and depression. Table 9.1 summarizes somatic disturbances and drugs that are most commonly associated with depressed[6,7] and manic states.[7,8]

The diversity of most of these conditions suggests that the depression is a *secondary* phenomenon that results from an interaction of psychosocial and biological elements. For instance, in Cushing's disease, while the elevated steroid levels secondary to pathology of the hypothalamo–pituitary–adrenal axis certainly dominate, especially in women, the profound changes in body image including acne, excessive hair, and obesity probably also contribute. Similarly in lupus erythematosus, an illness for which prognosis is difficult to assess, the demoralization associated with the uncertainty may be as instrumental to depression as are the side effects of the steroid therapy or the brain changes secondary to the disease process itself. In schizophrenia depression may arise in response to the recognition that the illness will run a chronic course, with devastating psychological and social consequences, but the blockade of brain catecholamine systems by neuroleptic drugs necessary to the treatment probably also increases the individual's vulnerability. As a last example, alcoholism—a common psychiatric cause of secondary depression—induces serious interpersonal and vocational losses, as well as exposing the person to the pharmacological depressant action of ethanol.

Mania may also result from the interaction of a variety of physical agents and illnesses, although here the genetic permission seems dominant.[9] Pathological denial of loss or disability can create behavioral conditions characterized by elation, hyperactivity, and impulsive action. Some individuals, for example, respond to cardiovascular illness with denial of hypomanic proportions, dangerously increasing their activities in the physical, sexual, and professional spheres.

Let us, by contrast, now consider those affective states that appear to be *primary* in that (1) the mood disturbance is the core clinical manifestation of the illness (2) the affective disorder chronologically precedes any other psychiatric disturbance, and (3) no associated illness or drug factors can be identified to help explain the profound despair or the inordinate elation experienced by the individual. On close inspection these primary dysfunctions are found to have all the attributes of a disease: significant morbidity and mortality, disruption of vegetative functions, autonomous course, and the usual requirement of biologically focused treatments to reverse them. Family history is often, though not invariably, positive in these primary affective states, with the pattern of familial distribution suggesting a genetic contribution to their etiology.[10] In origin they appear to be profoundly different from those affective states that are associated with somatic disturbance. Despite these differences the various primary and secondary forms of mood disorder share a remarkable similarity in clinical symptoms and signs. Furthermore, when one compares them by the nature and degree of functional impairment that is experienced, one is struck with their common features as syndromes rather than by their differences.

This relative homogeneity of the clinical syndromes found in mood disorders, despite considerable heterogeneity of the factors that interact to produce them, suggests that they are the manifestations of pathophysiological responses that have a final common pathway. In their fully developed forms the syndromes of mania and melancholia are characterized by disturbances of psychobiological function normally subserved by the limbic–diencephalic parts of the brain. Indices of biological change coexist with the behavioral dysfunction. From our previous review we may list some of them:

1. Disturbances in the sleep–wake cycle, including changes in sleep architecture such as a shortened rapid eye movement (REM) latency period[11]

2. Disturbances in neurophysiological indices of arousal, including cellular ion balance[12]

3. Disturbances in psychomotor activity which may progress to stupor[13,14]

4. Disturbances of mood, which in many persons exhibit a marked circadian variation[15]

5. Drive disturbances, particularly eating and sexual activity[16]

6. Changes in gratification and experience of pleasure[16]

7. Menstrual irregularities[17]

8. Abnormalities of biogenic amine metabolism in special subgroups, including 3-methoxy-4-hydroxy-phenylglycol excretion—the central norepinephrine metabolite—and 5-hydroxy-indole-acetic acid (5-HIAA), a metabolite of the serotonergic system[18,19]

9. Disturbance of the cortisol circadian rhythm with resistance to suppression by dexamethasone in about 50% of individuals[20]

10. Abnormalities of the brain–thyroid axis reflected as disturbances of circulating hormone levels and abnormal thyrotropin release upon challenge with thyrotropin-releasing hormone in approximately one-third of individuals[21]

Evidence suggests that in melancholia there exists a circumscribed core of psychobiological disturbance in arousal, appetitive, psychomotor, circadian, sleep, neuroendocrine, and pleasure functions, all the mechanisms for which are located in the midbrain. In manic states, also, many of the same functions are disturbed. Although clinically the deviation may appear to be in the opposite direction from that observed in melancholia, upon investigation some—e.g., sleep, arousal, and psychomotor functions—are found to be disturbed similarly.[12] Adopting the perspective that an individual together with the immediate environment constitutes a general system, these dysfunctions and the associated behavioral syndromes may be interpreted as representing a new dynamic steady state, a psychobiological deviation, functionally distinct from adaptive health and reached from diverse origins through common mediating mechanisms. *The new state is thus properly recognized as a function of the dynamic behavior of the system itself rather than the product of a single external variable.* This novel state of function is reversible, but while it exists the regulation of the usual interactive processes, both toward the larger environment and within the psychobiological self, is disturbed and the individual's optimum adaptation is compromised (Fig. 9.1).

Accepting this postulate, we now turn to a discussion of those factors that might interact to predispose an individual to the development of this dysfunctional state.

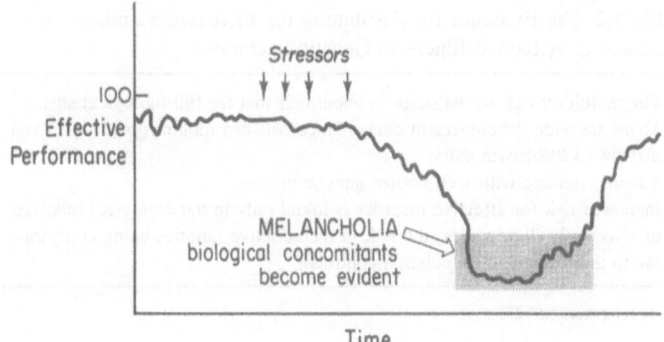

Figure 9.1 Diagrammatic conceptualization of melancholia: Stressors impinging upon a predisposed individual perturb performance until the dynamics of the system are such that a step function occurs. A novel set of behaviors emerges which is described as melancholia. Optimum adaptation is lost while this new quasisteady state persists.

The Predisposition to Disorder

In describing the factors that interact to provide permission for the development of mood disorder, we shall adopt the semantic clarification suggested by Winfree.[22] Items that appear relatively stable, such as family patterns of disease implying perhaps a genetic predisposition, we shall call parameters. Things that may change spontaneously and frequently, for example, attachment relationships with other individuals, we shall term variables. At times interaction with key environmental variables may so change the system that a new parameter of vulnerability becomes established; an example would be a destructive interpersonal experience surrounding the loss of a parent in childhood. Finally, of course, it is important to recognize that these definitions are relative and are a convenience. Parameters show stability relative to variables; it is not implied that only the latter change.

Genetic Vulnerability

Primary affective disorders are familial illnesses. The biological relatives of both unipolar and bipolar patients are at greater risk than the general population for developing affective illness[23]; the increase is of greatest significance in bipolar persons.[24] Familial is, of course, not synonymous with genetic, but four lines of evidence converge in attributing a major portion of the increased familial incidence to genetic factors (Table 9.2). The strongest is provided by adoption studies,[25,26] in which the incidence of depression in adult depressives adopted shortly after birth is correlated with the occurrence of depression in their biological, but not adoptive, parents. Gershon and associates[27] at the National Institute of Mental Health have estimated from their work that inheritance may contribute 50% of the variance in the pathogenesis of major affective disorder. This figure represents the mean estimate for all forms of primary affective illness, with the greatest weight contributed by the bipolar disorders.

The nature of the vulnerability transmitted by genetic factors in affective illness is

Table 9.2 The Evidence for Attributing the Increased Familial
Incidence of Affective Illness to Genetic Factors[a]

1. The morbidity risk for half-sibs is about half that for full biological sibs.
2. There are wide differences in concordance between monozygotic (70%) and
   dizygotic (20%) twin pairs.
3. There is linkage with well-known genetic markers.
4. Increased risk for affective disorder is found only in the *biological* relatives
   of affectively ill adoptees (the risk in the adoptive families being comparable to that of general population controls).

[a]From references 25–27.

unknown at present; it is probable that the specific dysfunction will vary among patient
groups. Subdivision of bipolar and unipolar depressions on the basis of family history
and treatment response potentially can help clarify the situation. Fieve has proposed
subdividing bipolar and unipolar depressions into six groups based upon positive or
negative family history of depression and mania.[28] Those with bipolar illness and a
positive family history for mania—compared to those with a negative family history
for affective illness—may have a more severe form of the disorder. Although the
incidence of suicide is greater, such individuals have a good response to lithium
carbonate treatment.

If such subdivisions emerge as valid, they may assist researchers in the identification of specific markers of genetic vulnerability, better specifying those at risk within
the pedigree of patients possessing the marker and perhaps even in the general population. Several biological indices are candidates to be considered as valid markers at the
moment, and the list is growing. A low level of 5-HIAA in the cerebrospinal fluid,[29]
an abnormal lithium ratio between red blood cells and the plasma,[30] platelet binding of
radioactively labeled imipramine,[31] and early induction of REM sleep by cholinergic
drug challenge[32] all appear possibly heritable traits: they seem to persist beyond the
period of illness in many persons who suffer mood disorder and exist in their "well"
relatives, but are not present in the general population.

Although such genetic markers will become increasingly important in the future,
alone they are merely an index of risk. By themselves they are probably neither
necessary nor sufficient to engender disease. Many sporadic (i.e., nonfamilial) forms
of both unipolar and bipolar illness exist. Furthermore, even when the role of heredity
is unequivocal, the expression of the genetic component can be confused by the
familial aspects of the disorder. The adopted child may emerge unscathed, while the
child who is reared in a family with primary affective illness is exposed to dual
disadvantage. Genetic vulnerability to affective illness is compounded by loss during
childhood through parental illness, suicide, separation, divorce, or more subtle forms
of affective deprivation secondary to disturbed interaction between the parents. Indeed,
assortative mating is common among those predisposed to mood disorder, and thus
children may be reared by two emotionally disturbed parents, with illnesses ranging
from personality and affective disorders to alcoholism.

Predisposing Temperament

The concept of temperament is an old one, referring originally to those mental or
physical features thought to reflect the individual mixing of the body humors. The term

was used in Kraepelin's time to describe the "constitutional" elements contributing to clinical disorder. Kraepelin himself considered four temperaments as the fundamental states in manic–depressive illness: the depressive temperament, the manic, the irritable, and the cyclothymic.[33]

In describing what are probably phenotypic variants of affective illness, the concept still has utility. Individuals first seen in the outpatient setting and followed for a considerable period of time frequently present with vague symptoms of mood disorder which fall short of the classical affective syndromes. The temperamentally dysthymic are often sluggish, have a negative self-concept, and despair easily. The hyperthymic seem always to be driven—pursuing various worthy and unworthy goals. Cyclothymic individuals are those "who constantly oscillate hither and thither between two opposite poles of mood, sometimes rejoicing to the skies, sometimes sad as death"—to quote Kraepelin's original description. All three groups are overrepresented in the families of those with major affective illness, especially bipolar illness.[9]

In such persons one is tempted to presume that the genetic potential to affective episodes is always subclinically active, to be triggered with ease into clinical illness by environmental challenge. Indeed, prospective follow-up studies at the University of Tennessee Mood Clinic suggest that all of these temperamental tendencies—including the predominantly hyperthymic—are at high risk for depression. In cyclothymic individuals, episodes of hypomania were frequently precipitated by tricyclic antidepressant treatment; the frequency was in fact comparable to that occurring in persons diagnosed as having manic–depressive psychosis (see Table 9.3), suggesting that there is more than a chance association between the two behavioral states.[34] Similarly shortened REM latency, the possible trait marker found in major depressive illness, has recently been reported to be present in all three temperaments.[35] Hence it is our belief that the concept of temperament reflecting inherited subclinical factors remains a useful one for the clinician weighing the parameters that interact to permit affective illness. Further progress of the research into trait markers, however, will aid in assessing the validity of this supposition.

Table 9.3 The Increased Risk for Affective Episodes in Those of Cyclothymic Temperament Compared to Controls in a Two-Year Prospective Study

Clinical affective episode	Cyclothymic temperament ($N = 60$)		Control group ($N = 50$)	
	Number	Percent	Number	Percent
Drug-free course[b]				
Depression	18	30	2	4
Hypomania	9	15	0	0
Mania	3	5	0	0
Course on tricyclic drugs[c]				
Hypomania	11	44	0	0

[a]From reference 34.

[b]Significant cumulative increased risk for affective illness over control group ($p < 0.001$).

[c]When treated with tricyclic drugs those of cyclothymic temperament developed mania at a similar rate to individuals with known bipolar illness [44% and 35% respectively, a significant elevation over controls ($p < 0.01$)].

The Modulating Influence of Age and Sex

The incidence of unipolar depressive illness and the frequency of episodes increase with age.[36] Bipolar disorder, which typically begins before age 30, changes in quality with advancing years; the frequency and length of the depressive episodes rise relative to those of excitement.[36] All varieties of depressive disturbance are more common in women.[37] This is true in virtually all cultures on which reliable data are available. Only in manic–depressive psychosis is the prevalence ratio approximately equal between men and women.

In our culture these age and sex differences are frequently ascribed to social and psychological factors. It is argued that elderly persons have a greater exposure to personal loss at a time of declining social support. Similarly, women may be predisposed by a passive and dependent social role to separation difficulties, especially in midlife. While undoubtedly these sociocultural elements are important, they also appear to be intimately entwined with biological variables. The activity of monoamine oxidase enzymes, which it will be recalled are crucial catabolic agents in the life cycle of the monoamine neurotransmitters, increases with age. Serum concentration is also higher in women than in men, regardless of age. Such increased activity indicates a more rapid breakdown of monoamine transmitters and presumably their faster removal from the synaptic region. Evidence suggests that the production of monoamines may also decline with age, compounding the situation and conceivably increasing the individual's vulnerability to biogenic amine imbalance under challenge.[38]

Thyroid hormones, particularly triiodothyronine, confer an advantage on some patients receiving antidepressant medication.[39] This effect appears to predominate in women. Thyroid activity measures are in general lower in women than in men throughout life—in the normal population and in depressed patients.[40]

Some of these parameters specific to the vulnerability of women are found in Table 9.4. Depressive symptoms have also been associated with the premenstrual,[41] postpartum,[42] and menopausal[43] periods, as well as with contraceptive use,[44] suggesting that the psychoendocrine changes occurring at these stages of the female life cycle interact with other permissive parameters to increase the woman's specific vulnerability.

Table 9.4 Modulating Biological Factors in Depression, Especially Relevant to Increased Risk in Women

1. Higher levels of brain monoamine oxidase, the enzyme that breaks down monoamine neurotransmitters
2. Precarious thyroid economy
3. Two X chromosomes (relevant if X-linkage is the mode of transmission of a genetic subtype of bipolar illness)
4. Occurrence of postpartum episodes
5. Premenstrual precipitation
6. The use of steroidal contraceptives

Physical Illness and Its Treatment

Earlier in this chapter, when we set forth the evidence that affective illness is a final common path of dysfunction, we reviewed many of the physical illnesses associated with melancholia and mania (see Table 9.1). We also noted the inevitable intertwining that occurs between the biological demands made by the illness itself and the reaction of the patient to his or her new psychosocial predicament. Hence medical illness can so complicate an individual's management of his life and his physiology that it may emerge as both a predisposing parameter and a precipitating variable in the initiation of mood disturbance. Metabolic disturbance, insomnia, and pain interact with concerns for narcissistic integrity, loss of control of bodily function, and loss of important skills or attachment. For even the most resilient individuals the stressors can be overwhelming. A life-threatening illness may precipitate an acute grief reaction—for the self—or a life-threatening denial. Should the illness become chronic, a whole new set of adaptive templates may be required to accommodate the new life style that is imposed. The role that physical illness may play is thus extraordinarily complex.

However, we may also learn more about the mechanisms that underpin mood especially by paying close attention to those medical treatments that could potentially have an impact on mental function. Reserpine remains a classic example. Of those hypertensive individuals receiving doses of 0.5 milligrams per day or higher, 15% became depressed.[45] However, close inspection revealed that two-thirds of the reported cases were pseudodepressions, with the dominant change being sedation. Most of these individuals recovered as soon as the reserpine was discontinued. True melancholia was restricted to approximately the remaining one-third, or 5% of the total. These individuals continued upon an autonomous course despite withdrawal of the offending chemical. This suggested that the risk of developing a "true" depression on reserpine is actually less than the 15% lifetime risk for affective illness in the general population, and in fact, those who did develop the disturbance either had previously been depressed or had a family history of affective illness.

These findings are extremely important for the biogenic amine theories of affective disorder, for they suggest that the mere depletion of biogenic amines with reserpine is not a sufficient cause for depressive illness. This conclusion is further supported by the failure to induce a full depressive syndrome in normal persons with alpha-methyl-paratyrosine or para-chlor-phenylalanine, selective inhibitors of norepinephrine and serotonin synthesis respectively.[46] We may conclude that changes in biogenic amine balance alone have little power, but in individuals with an affective predisposition the interaction may initiate clinical disorder.

What then of the interaction of chemical and psychosocial stressors, an important issue for those physically ill persons who receive drugs with a potential for creating monoamine imbalance? For example, if our concepts of psychobiological homeostasis have validity, then bereaved individuals exposed to reserpine will be more likely to develop melancholia than those not receiving it. This would be predicted because mobilization of the adrenergic regulatory mechanisms, necessary for coping with the stress induced by the loss of attachment, will be compromised by reserpine's pharmacological interference. This specific conjecture has not been formally put to the test in the clinical setting, for obvious ethical reasons. However, in modified form the experiment has been undertaken in the Wisconsin Primate Laboratory. Juvenile-age monkeys exposed to a separation paradigm that had not previously produced depressive behavior developed the syndrome when the catecholamine-depleting drug alpha-methyl-paratyrosine was administered concomitantly.[47] Similarly, in the clinic there is now some evidence that a combination of catecholamine-depleting antihypertensive agents and multiple losses of immediate family members is associated with chronicity in late-onset depressive illness.[48]

Loss of Attachment in Childhood

It would appear from primate studies that attachment is a primary drive for animals such as ourselves. Total failure of attachment bonds, by whatever cause, results in gross psychopathology. Imperfect resolution of the attachment–separation developmental complex, although not as catastrophic, impairs self-regard and psychobiological autonomy. However, as we learned from Engel and Schmale's studies of Monica,[49] even overwhelming environmental challenge produces a form of adaptation. Monica's conservation withdrawal was adaptive to the minimal nurturing she received. This passive sleeplike state associated with virtually no motor activity and cessation of gastric secretion became associated with subjective feelings of helplessness and hopelessness as conscious recognition of the vulnerability of the self emerged.

Psychoanalytic theorists have long suggested that a depriving early environment, which impedes or prevents mastery of the experiences surrounding attachment and loss, predisposes an individual to depression in adulthood. Bowlby, in a recent formulation,[50] broadens the concept a little to postulate that early separation and unsubstituted loss may predispose an individual to a spectrum of adult personality disturbance characterized by dysphoria, antisocial behavior, and possibly suicide.

The epidemiological data generally lack specificity, but several interesting associations have been reported which lend some support to Bowlby's thesis:

1. From a study of persons with unipolar melancholic illness, Perris[51] has reported that those individuals experiencing loss in childhood on the average develop their disorders 10 years earlier than those without such loss.
2. Beck and co-workers[52] have reported that the high Beck depression inventory scores in adult depression tend to be positively correlated with a history of childhood loss.
3. Other investigators[53,54] have found suicide attempts to be more common in those depressed subjects who have experienced early separation from parenting persons.
4. In a heterogeneous population of ambulatory depressives, studies at the University of Tennessee[55,56] have shown that developmental object loss is correlated with the adult occurrence of "unstable" characterological features of immaturity, hostile dependency, manipulativeness, impulsiveness, and low threshold for alcohol and drug abuse.

These findings suggest that, while clearly not causing depressive illness, loss of interaction with important persons during development can modify the expression of the adult condition, becoming a potentially significant parameter in reducing the threshold for mood disorder. It may predispose in particular to the development of characterological constellations that foster affective illness.

Predisposing Character Traits*

In human beings the basic psychobiological process of attachment and separation is almost infinitely modified by abstract conceptual thought. Our ability to adapt to—and to modify in our own interest—the environment around us is a function of our particularly complex and vigilant central nervous system, which integrates, modulates, and records every detail of the interaction. During development we learn to supplement with attachments to others the exclusive caring provided initially by parental figures. Gradually we come to master our own environment, developing idiosyncratic adaptive styles that in adult life provide the individual distinction we term character.

In most individuals these personal behavioral templates continue to evolve in complexity as we negotiate the inevitable challenges of the life cycle[2]—leaving home, developing an identity as a person with marketable skills, illness, sexual attachment and loss, marriage, children, and the death of those close to us. There are occasions, however, depending upon physical health, the social supports available, and the specific nature of the challenge, when we may falter in this interaction and reexperience those earlier feelings (characterized by Schmale and Engel[49] as helplessness and hopelessness) which were associated with a subjective loss of attachment and mastery. Such feelings may be reported to others as a transient depression of mood. However, usually the individual with normal self-regard, if unfettered by other predisposing factors, struggles against these feelings, finds other adaptive modes, and does not give up despite the urge to conserve and withdraw.[2]

A fundamental tenet of psychodynamic theory is that our ability to cope successfully in relation to the demands of the environment is profoundly influenced by the adaptive styles we embrace. Most of us have preferred ways of coping to which we return at times of greatest challenge and complexity; these are the dominant character traits that distinguish us as adults. How important are these characterological parameters in the development of mood disorders? Are there *specific* adaptive styles that predispose an individual to affective illness?

Objective studies of the question are exceedingly difficult to conduct; indeed, probably none of those existing offers a satisfactory approach.[57,58] Adaptive styles are patterns of behavior which emerge over time and in relation to challenge. Hence ideally studies should be longitudinal and prospective, evaluating groups of individuals at risk prior to the development of illness. Most psychodynamic clinical observation, while longitudinal, is subjective and restricted to individuals who already have affective illness; most "objective" studies quantify the behavioral style by paper and pencil tests but at only one point in time, and again are largely retrospective. Thus in formulating our conclusions here, we have drawn upon both sources of information.

The clinician routinely gathers information about premorbid personality by asking

---

* The semantic distinction between character trait and temperament may be considered by some to be a fine one. Here we use the term *character trait* as a reference to a behavioral attribute that particularly distinguishes an individual. A component of learning is implied in the development of such traits, while for purposes of discussion we do not consider this to be the case with temperament. Personality is reserved as the term to describe the totality of an individual's behavior, emotional attitudes, and habits, and thus includes both traits of character and temperament.

about past history. Together with the longitudinal observations made during illness and the process of recovery, this data base has led to specific psychodynamic formulation of predisposing character traits in depression. They are most easily understood within the dynamic paradigm of attachment and loss.[59] The hypothesis is essentially as follows: Having never satisfactorily resolved the psychobiological issues surrounding attachment and separation, depression-prone persons remain extremely sensitive to loss or perceived loss—of people, of animals, of their own philosophical goals, of friendship, even of material objects. Frequently, in defending against this contingency and its associated feelings of helplessness and hopelessness, such persons have adopted rigid adaptive styles. Some of these styles or rescue operations work very well for many years, but frequently at middle age, a special combination of events disrupts the precarious equilibrium and with it their self-regard. Adopting this perspective, three overlapping clinical clusters of character traits seem prominent in those depression prone.

Dependence upon a Dominant Individual

The most striking cluster is that through which *adaptation is achieved by dependence upon a dominant individual.* Here the basic rescue strategy has been to successively transfer the unresolved issues surrounding attachment and separation from the parenting figures to others in the individual's life who appear competent and capable. Thus no true self-reliance develops during the maturing years, and the individual enters adult life with a constricted series of adaptive options. Such individuals may find it difficult to enter the usual commerce of adult life, for their own self-esteem is tied too closely to the opinions of those around them. They appear shy and introverted. Stable adaptation is not impossible, however, especially if the relationship with the dominant figure who maintains a care-taking role has mutually satisfying elements.

Clinical experience suggest that this form of adaptation in our culture is found most commonly in women. Very often such individuals are good mothers of young children, for in the infant they are able to vicariously invest their own need for dependence and closeness. However, as the children grow, they may have greater difficulty than usual allowing the offspring to express their autonomous strivings. Thus toward middle age the adaptive style may begin to crack as the dedication and investment in the affairs of the family bring diminishing rewards. Without support, a sense of helplessness and then hopelessness ensues. Anger may erupt—driven by a sense of never having been truly provided for or cared about—with mutual withdrawal and further disruption of the social reinforcement systems. Alternatively some may develop increasing hypochondriacal complaints, or use genuine physical disability to frequently consult their physician and thus transiently reunite with a dominant, caring person. Prescribed medications and also alcohol can be abused, merely compounding the problem.

Obsessional Traits

Another widely hypothesized predepressive style is *that in which obsessional traits pervade the character.* Here one may postulate that when there is an inconsistency in the available caring figures during the maturational years, a potential adaptation is to reduce to a minimum the number of those situations in which the difficult challenge of resolving separation anxiety is likely to occur. Adaptation becomes the control of the day-to-day details of life. Basic to the strategy is the restriction of novelty; planning becomes paramount, and those who bring surprises are greeted with a hostile response. When such a mechanism becomes a predominant character trait, one finds little spontaneity—rather a meticulous and very reliable individual. Thus frequently all goes well in early adult life.

However, that the individual has had little opportunity to develop a fundamental trust of others rapidly becomes apparent when the adaptive style is challenged by an increased number of events that the individual can no longer control. Increased obsessional behavior, overt suspicion, and even paranoia emerge, increasing the social isolation. To the bystander, depressive illness in such individuals may seem to emerge de novo: a highly successful, conscientious individual who is suddenly destroyed by suspicion or malevolent self-doubt. (The lawyer in our opening chapter may be taken as an example of such.)

Pseudoindependence

The aggressive, pseudoindependent individual falls into a third, less easily defined category of vulnerability which is possibly related to bipolar episodes. To all outward appearances, such individuals are the cultural stereotype of the independent person, although they may avoid intimate contact with others. They usually have high aspirations for themselves and similar standards for others; when tied to a well-structured philosophical ideal based upon social or religious beliefs, they may enjoy serving others and usually do so well. In other instances their energy and excessive demands suggest a driven, almost hypomanic adjustment. Frequently they have broken away from their primary family early in life and set up, through their own effort and industry, an independent life of their own. They tend to make people dependent upon them, perhaps thereby gaining some vicarious gratification. However, not far beneath the surface and usually within their awareness, they frequently yearn for a close relationship with a caring individual, with someone, in fact, who fits the ego ideal for which they themselves reach.

Psychometric Studies

These adaptive styles are not specific for affective illness. Excessive dependency, for example, is found in many disorders. It is the interaction of such characterological parameters with the other predisposing elements and the precipitating variables that presumably provides the cumulative and perhaps specific vulnerability.

Recent studies have renewed efforts to quantify these formulations, and some interesting information has emerged. Several psychometric studies have lent support to the hypothesis that the premorbid personality of those with late-onset depression is self-critical, conscientious, and hard working; the individual responds to adult losses in a self-punitive and self-denigrating way.[60-62] Preexisting obsessional traits have long been clinically associated with melancholia, but this is not confirmed generally by the psychometric studies. Indeed, in the National Institute of Mental Health Collaborative Study on the psychobiology of depression,[62] obsessionalism was predominantly a feature of those with bipolar illness, suggesting that the extroverted, ambitious, and driven individual may indeed have many obsessional characteristics, being basically duty bound and work addicted. Some special character traits of the depressive do appear in such studies; predominantly they are those of introversion, a lack of self-confidence, nonassertiveness, lack of social adroitness, and dependence.[62,63] Most of the information comes from the history of persons after recovery from illness, however, and thus some of these findings may reflect a character constellation secondary to or magnified by the illness itself.[64] Indeed, Metcalfe has argued that it is a loss of resilience rather than a set of premorbid traits that best characterizes the depression-prone person.[65]

Character and Adaptive Style

Predisposition to disease is not static but varies at any one moment of the life cycle as a function of the dynamic interplay among the predisposing parameters and their cumulative feedback from environmental interaction. An individual's adaptation is strongly influenced by characterological style. These learned paradigms have a major influence in the appraisal of psychosocial stimuli. For example, if the environment is seen as a potentially hostile place that must be controlled, then even inconsequential events will become the focus of a great deal of energy. This increases the general strain imposed upon the individual by *all* events. Furthermore, if the cognitive schemata or templates involved lead to the personal conclusion that mastery in certain social situations is difficult or impossible, then such self-evaluation will provide reinforcement to avoid all similar situations. The result of this cycle is social withdrawal resulting in an even greater decrement of those skills with which to manage the next social challenge. Increasing behavioral withdrawal is the natural outcome. It may be such a dynamic series of events that leads to those introverted character traits now thought to be most consistent as premorbid attributes in primary unipolar depressed persons.[62,63]

Precipitating Factors That Initiate the Path to Dysfunction

In our preceding discussion we have suggested that a predisposition to affective illness exists as a set of continuous and reasonably stable variables which we have termed parameters. The predisposing power of these parameters varies between individuals, being dependent upon psychobiological characteristics both genetic and acquired and further modified by factors such as sex, the aging process, illness, and adaptive styles. Although these factors together are formative in the pathogenesis of an episode of clinical illness, the syndrome itself usually emerges in close temporal association with environmental challenge.

The nature of these challenges varies broadly, of course; changes in the daily photoperiod or temperature may be important for those predisposed to seasonal depressions, for example. Similarly, rapidly crossing time zones may precipitate mania or pharmacological agents and illness again may serve as precipitants as well as predisposing parameters. The inherent complexity of the intertwining of person, process, and event makes an absolute classification misleading. However, the most common variables, at least in the precipitation of primary depressions, are psychosocial events, and it is upon these that we focus in this section.

Psychosocial Variables as Precipitants

Of those persons predisposed to melancholia, only about one-third when faced with "exit" life events experience clinical depression. Thus even those at risk do not develop mood disorder every time they sustain a loss; ultimately, whether decompensation occurs is a function of the temporal clustering of the stressors, their qualitative interpretation, and an individual's orchestration of the interaction required to moderate them.

Loss of Attachment

Loss of attachment to individuals, goals, or ideals is a central theme of the challenge which everybody finds difficult, but it is particularly disorganizing to those who are depression prone. As we have discussed previously, unsatisfactory resolution of similar early experience may play a major role here, having helped create the "faulty templates" upon which an individual must rely in developing response appropriate to the loss.

The Importance of Social Supports

In such individuals the extent and quality of social support thus becomes an important factor in determining the outcome of uncomplicated bereavement. In the well-controlled studies on grief by Clayton,[66] depression emerged more frequently in those without grown children to offer support in the postbereavement period than in those with a strong family history (and therefore presumed genetic predisposition) of affective disorder. Similarly, research now suggests that the syndrome of anaclitic depression, originally described by Rene Spitz,[67] occurs in only 15% of children when they are permanently separated from their mothers, if sensitive substitute mothering is available.

To some extent in children but particularly in adults, the quality and degree of social support received by individuals is a function of the ability they have to interact with those around them. Depression-prone persons and especially the elderly are therefore placed in double jeopardy.[68] People who possess a narrow repertoire of social and vocational activities, as is frequently the case in those who have character traits that predispose them to depression, find themselves unable to engage in activities that provide adequate sources of substitute gratification for their personal and social needs. The much-needed reinforcement required to maintain normal behavior just does not occur (a point we return to a little later), except as it is distorted in their dependent need for physicians, hospitals, and treatments. In the absence of adequate social support, the sick or depressed role may then become a way of life—a way of assuring attention, sympathy, and interpersonal support.[69]

Loss of Symbolic Attachment and Loss of Control

The loss of symbolic attachment may also at times act as a potent force in determining whether a depressive syndrome will develop. This is particularly true when one goal in life has precluded the development of broad socially adaptive skills. For instance, a young woman whose exclusive life dream was to be an accomplished pianist became severely depressed when early rheumatoid arthritis began affecting her hands. The importance of understanding the personal meaning to an individual of the challenges they find difficult to manage cannot be overemphasized. Without such attention to psychodynamic detail, the true precipitants of an episode of illness may never be elucidated. In exploring these intrapsychic dimensions that embellish the perceived stressor, it is important to recognize that events that may appear superficially advantageous in fact also may be overwhelming. Hence an individual who is promoted may find the new position attractive but threatening. An initially poor performance then results in criticism and loss of self-esteem, quite in contrast to the success achieved in a previous less important post. Frequently stressors of this variety are related to a *loss of control* of environmental interaction, either actual or feared. Obsessional character traits make an individual particularly vulnerable to interpreting novel life situations in this way. The stress induced results from the adaptive demand imposed by a significant departure from routine with little relationship to whether the quality of the change is potentially noxious.

Frequently, of course, for those who are particularly vulnerable, loss of control and loss of attachment go hand in hand, each situation magnifying the interactional demands of the one that preceded it. For example, young women with both unipolar depression and a positive family history of alcoholism appear particularly vulnerable to this downward spiral. When finally they seek help, their chaotic lives frequently seem governed entirely by emotional adversity to which they have developed a set of capricious character traits. As templates these traits are sadly maladaptive in the larger world. By contrast, those with "superego strength" and controlling personalities with apparent independence in youth develop depression later in the life cycle even at times when there is a strong family history for depressive illness.[61] In this group the onset of depression is typically after the age of 40. In these two groups familial–genetic, age, social, and characterological parameters are seen as interacting to determine a clear-cut difference in the threshold and the time in the life cycle for the depressive decompensation.

The Number of Events in Unit Time

The number of events demanding attention per unit of time will also have a major impact on the way an individual handles any one of those events. The speed with which response is required threatens the control that can be maintained in orchestrating the interaction to one's best advantage. Not surprisingly, therefore, the available evidence favors the view that the depressant impact of a personal loss is considerably increased by its coincidence with other life events that also demand resolution. From the work of Paykel and associates,[70] we know that depressives report three to four times the number of life events per unit time than do normal individuals. These studies have now been replicated in other cultures.[71] Obviously this cumulative excess of stressors may not be real; however, the perception that life is overwhelming becomes an important factor in threatening the psychobiological homeostasis.

Precipitating Factors in Mania and Hypomania

By comparison with melancholia, there has been little research conducted on the precipitating psychosocial variables specific to mania and hypomania. This is largely because mania is not an entity distinct from depressive illness—at least only very rarely. Usually it is closely associated with a depressive episode and the continuum from one phase of the illness to the next contains many mixed behavioral states. Hence in seeking specific precipitants investigators may not be asking a question critical to a better understanding of the disorder. Where such studies have been undertaken, most evidence suggests that the challenge of life change plays a role similar to that in depressive disturbance. Again, loss of attachment and control threatens to lower self-esteem, and a manic response results because of the genetic predisposition of the individual rather than any specific quality of the environmental challenge. Ambelas, in a recent controlled British study[72] of a group of manic–depressive patients, carefully excluded any life event that might have been precipitated by the manic excitement itself and found that an excessive number of environmental demands preceded the illness; findings similar to those of Paykel in depressives.

Psychoanalytic theory has for long held the view that mania is a "mask" for depression—a defensive attempt to ward off the pain of loss. Indeed, French psychiatrists have anecdotally described "*le deuil maniaque*" (or maniacal grief), where a widow exhibits inappropriate elation, hyperactivity, expansive behavior, and lack of insight.[73] Controlled studies of this phenomenon are needed, however, if it is to be understood within the context of the hypomanic or manic states of bipolar illness.

In fact, the best-known precipitants of hypomania per se (and to a lesser extent of mania) are drugs and physiological procedures that increase the central level of catecholamines. The noradrenergic tricyclic drugs, the monoamine oxidase inhibitors, ephedrine, L-dopa, thyroid hormones, amphetamine-like drugs, REM deprivation, and total sleep deprivation[74] all most consistently produce a switch from depression to hypomania. Conversion of an individual from a euthymic state into hypomania rarely occurs. Thus "pharmacological switching" is largely limited to individuals with a bipolar diathesis—measured as a function of personal or family history of mania. This again emphasizes that the relationship between melancholia and mania is physiologically determined and psychosocial factors play little if any specific role in "selecting" the manic syndrome. Parenthetically, it is worthy of note that fusaric acid—which inhibits dopamine-beta-hydroxylase and thereby increases central dopamine levels—transforms hypomania and mania into a florid manic psychosis.[75] From this evidence and the effectiveness of dopaminergic-blocking agents in mania, we may tentatively conclude that dopamine metabolism contributes specifically to the psychotic symptoms in florid mania as well as to the increased motor activity.

Intermediary Mechanisms to Adaptive Dysfunction

Appraisal, Response, and Pathway Initiation

We have emphasized a number of times, in discussing human response to psychosocial challenge, the importance of the appraisal process (see summary of how complex systems behave at the end of Chapter 8). Human beings are unique in the animal kingdom for the psychological abstractions that they bring to this task. Whether the event is disorganizing of psychobiological equilibrium depends in large part upon whether it is threatening to attachment, control, or similar parameters of personal integrity. These are now familiar concepts. It is this appraisal that constitutes the first intermediary mechanism between challenge and the common path to adaptive dysfunction.

Whether the challenge is familiar or unfamiliar, mild or severe, a single event or one rapidly replicated, manageable or not by established coping strategies—all these facets are appraised against established templates (spatiotemporal norms) of interactional behavior. Much of this effort is cognitive and highly visible to the individual, some of it is psychologically preconscious or unconscious. The appraisal process provides a gating mechanism to psychophysiological arousal. During a smoothly managed encounter in which no threat is perceived and coping easily sustains homeostasis, subjectively and objectively the event may leave little psychobiological record. Changes perceived as threatening, however, lead to subjective (i.e., emotional) and behavioral arousal with concomitant change in biological function. The innate ability of individuals to appraise and to adapt varies. Over the long run this is a function of the ability to see patterns in the environment and to learn from them. It is a function of motor response, of long-term memory, of the parameters of affective lability genetically determined. During maturation these fundamental biological factors interweave with a facilitating environment to maximize learning and adaptation and thus successful psychobiological coping.

Acutely, however, the intermediary mechanisms that underpin an individual's ability to evaluate and respond to environmental opportunity may be transiently impaired for reasons entirely independent of the event itself. Disturbance in the functional level of biogenic amines, endocrine systems, neuronal excitability, and receptor sensitivity are all examples of changes which may produce such impairment. Being tightly interconnected subsystems, these intermediary processes influence each other and the environmental interaction, so any sustained disturbance in one element will potentially drive change in those other elements of the system with which it is closely interlocked. Given sufficient perturbation, whatever the origin, a synchronous change in behavior of the dominant subsystems may occur, initiating the pathway to a new psychobiological equilibrium. It is within dimensions such as these that affective illness can emerge.

Biological Mediation of the Arousal Response

In Chapter 6 we reviewed the mechanisms that are involved in the mediation of the arousal response, in particular those changes encountered in the peripheral, neuroendocrine, and autonomous nervous systems. The responses are remarkably similar in man, primate, dog, cat, rat, and most other higher animals that have been studied. It has also been found in laboratory experiments that if an animal can establish control over the interaction with the environment the neuroendocrine arousal is reduced, emphasizing the feedback relationships between event, appraisal, and response which we discussed earlier.

When perturbing events persist or escape from behavioral control, certain general characteristics of biological arousal are apparent. Peripheral endocrine indices rise, reflecting an increased metabolic turnover of steroid, thyroid, and adrenergic hormones. Prolactin and growth hormone release occurs. Neurophysiologic measures indicating arousal such as increased heart rate, blood pressure, galvanic skin response, and EEG changes also are found. Subjectively the individual reports tension and hypervigilance. These changes are undoubtedly adaptive, especially within the short time frame. Frankenhaeuser,[76] for example, has shown that individual differences in mobilization of the peripheral adrenergic mechanisms under stress correlate with performance. Mason[77] offers substantial evidence in primates for a temporal organization of endocrine response which serves to maximize the chances of retrieving homeostasis (Fig. 9.2). Initially the normal circadian oscillation in these physiological variables is preserved. In the daily steroid rhythm one finds a general elevation of levels, but the basic temporal architecture remains intact. With continued perturbation, however, this curve begins to flatten and shift in its phase relationship with other cycles.

Within the central nervous system itself, animal studies provide most of our information regarding the stress response. A stress-provoking situation for rats that is easy to use in the laboratory is a short swim in cold water. It has been shown that this treatment lowers brain norepinephrine, increases the reuptake of labeled norepinephrine by the neuron, and produces a hypoactive behavioral syndrome for as long as 24 hours following the test. Also, in rats that swim to exhaustion brain norepinephrine levels were reduced by 20% and brain serotonin was elevated. The findings are interesting in light of the recent evidence suggesting that serotonergic neurons potentiate adrenergic systems in brain (see Chapter 7). The time courses of these alterations differed for the two transmitters: the increase in serotonin found immediately after the swim exercise had returned to normal levels within 2 hours; norepinephrine on the other hand, had not returned to control levels after 6 hours.[78]

In other laboratory situations with animals, cats developed a reduction of brain norepinephrine under challenge in both aversive and pleasurable situations, suggesting that the change is a correlate of heightened arousal or "emotionality."[79] In rats yoked experimentally, brain norepinephrine (in the hypothalamus and cortex) actually tended to increase in the animal that was given control over the number of shocks received, whereas in the yoked rats that could neither escape nor control the shock, the levels of brain norepinephrine consistently fell.[80] These data are consistent with those of Frankenhaeuser[76] in human subjects, which indicated that changes in the pattern of pe-

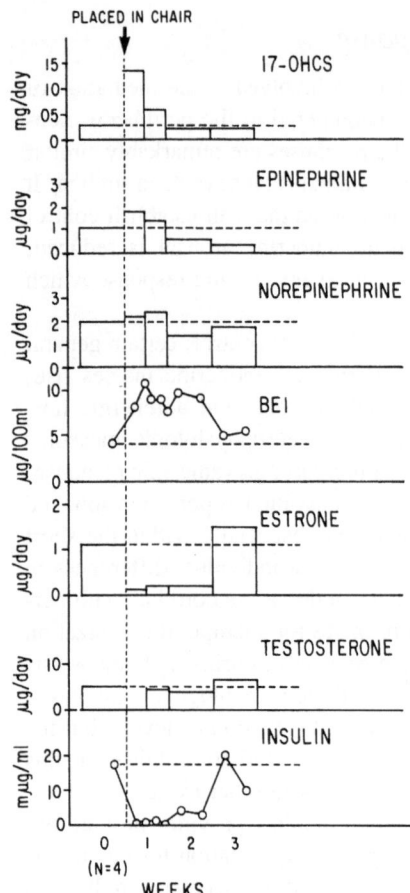

Figure 9.2 Orchestrated adaptation to restraint in a primate. The multiple hormone responses to being restrained in a chair have a characteristic pattern. Initially adaptation is mediated by increased levels of 17-hydroxy-corticosteroids and the catecholamines. Sex hormones and insulin are virtually suppressed and rise later. Thyroid mobilization falls in the middle range. (After Mason.[77])

ripheral adrenergic arousal occurred as psychological control over the experimental situation was achieved.

In general, the animal experiments, reviewed by Anisman and Zackarko,[81] support the proposition that noradrenergic metabolic turnover increases in brain under stress but that as challenge continues the functional units of norepinephrine available decrease. This change is reflected in hypoactive behavior, a syndrome that can be reversed by intraventricular norepinephrine. There is some evidence that the reduction in available neurotransmitter is related to increased neuronal uptake. In human studies of MHPG excretion, Sweeney and colleagues have reported a highly significant relationship within individual analyses between levels of the metabolite in the urine and state anxiety,[82] a finding that one may cautiously attribute to an increase in norepinephrine turnover similar to that reported in animal studies.

The Importance of the Diencephalon

The limbic structures, and particularly the hypothalamus, are the loci for those neurochemical events that mediate the stress response. Changes in the balance of biogenic amines in the hypothalamus regulate the release of the trophic hormones which orchestrate the peripheral endocrine response. There is also interaction of the released hormones in the periphery (e.g., steroids increase the conversion of norepinephrine to epinephrine). Ultimately they feed back upon the brain itself to influence biogenic amine activity in the hypothalamus—in a now familiar continuous cycle of behavior and response.

The many signs of diencephalic disturbance seen in depression and mania may reflect an abnormal interaction of these mediating mechanisms with the biological correlates of those parameters that predispose to affective illness. The disorganized neurophysiological hyperarousal indicating a loss of the usual balance among biogenic amine, electrolyte, and endocrine interrelationships then replaces the usually smoothly orchestrated adaptive response to perturbation. In such a deregulation a disturbance of the function of the reinforcement systems in the diencephalon becomes of particular importance. As reviewed earlier much evidence from animal studies now suggests that catecholamines, indoleamines, and acetylcholine as well as peptides play significant roles in regulating behavior through the dual mechanisms of the medial forebrain bundle (MFB) and the periventricular system (PVS), the neuroanatomical substrates of the "reward" and "punishment" centers respectively.[83,84]

Evidence from this work with animals suggests that lesions that interfere with the anatomical or chemical integrity of the reinforcement system in the diencephalon impair the ability of the organism to respond to environmental reinforcers. Therefore, whatever the nature of the events that mobilize depressive behaviors, their locus of interaction will probably involve the diencephalic centers of reinforcement.

An Integrative Model of Depression

To conceive of the syndrome of depression as the outcome of a single set of physiochemical variables is obviously an oversimplication. Depression cannot be equated with imbalance in one or another class of neurotransmitters, endocrine messengers, or disordered electrolyte metabolism. It appears that a potentially reversible neurophysiological state of hyperarousal—which may be determined, for example, by impaired monoaminergic transmission or intraneuronal sodium accumulation—occurs concurrently with the experience of frustrating environmental events that signal intense turmoil, impending decompensation, and hopelessness. The net result is nonrelatedness and anhedonia, with the concomitant failure in vegetative and psychomotor functions being experienced by the individual as additional evidence supporting the validity of such negative self-perception. Depression, then, is conceptualized as the feedback interactions of three sets of variables at chemical, experiential, and behavioral levels—with the diencephalon serving as the field of interaction. These concepts are schematically presented in Fig. 9.3.

The MFB and the PVS, which represent the anatomical substrates of reinforcement, have established feedback connections with other systems in the brain. Functional impairments in one system, then, could result in functional shifts in one or more of these systems. For illustrative purposes, and at the expense of gross oversimplifica-

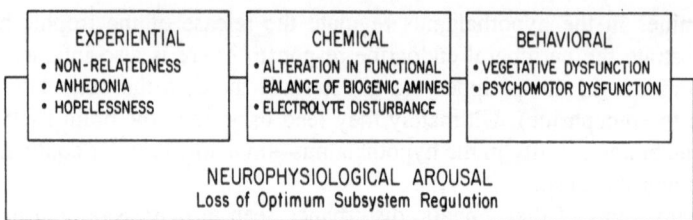

Figure 9.3 Melancholia conceptualized at experiential, chemical, behavioral, and neurophysiological levels.

tion, it would be instructive to consider the following three systems which are intimately related to the "reinforcement" (MFB–PVS) system: the "stress" (hypothalamopituitary) system, the "arousal" (reticular activating) system, and the "psychomotor" (pyramidal–extrapyramidal) system. All four systems, with their feedback connections, are diagrammed in Fig. 9.4.

Melancholic and manic behaviors, according to our scheme, result from a failure in the homeostatic mechanisms that maintain these systems in negative feedback—when physiocochemical alterations in these systems produce increasing levels of positive feedback in the reinforcement system. Stress or frustration beyond the coping ability of the individual—together with their psychic (anxiety, hopelessness) and neuroendocrine (increased cortisol, sodium retention) correlates—produce heightened arousal and thereby disrupt the functional integrity of the reinforcement system. The resultant decline in vegetative and psychomotor functions and the subjective perception of losing control to an impending state of decompensation serve as novel sources of stress. Arousal is further increased and coping mechanisms continue to decline. Thus a vicious cycle is established of more arousal, more hopelessness, and more evidence of psychomotor deficits. When affective illness is conceptualized within a dynamic system such as this, the controversy regarding whether altered catecholamine metabolism is a cause or effect of depression can be easily resolved. Impaired noradrenergic transmission would contribute to these diencephalic reinforcement deficits, regardless

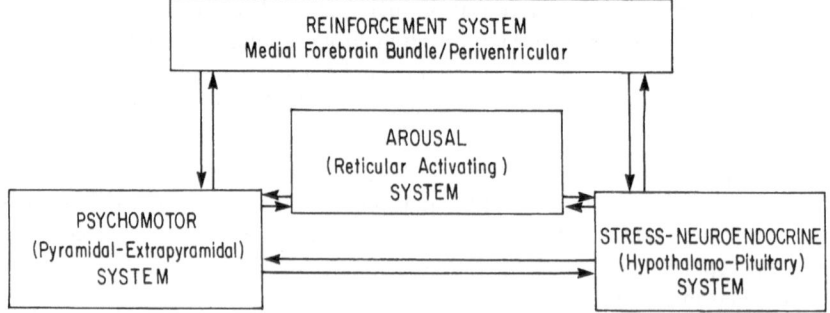

Figure 9.4 Hypothesized relationship of neurophysiological systems involved in mood disorders. Disturbance of the usual balance between these systems in melancholia is experienced as a disorganized hyperarousal and loss of meaningful interaction with the environment.

of whether the metabolic changes are primary or secondary. In other words biogenic amine disturbances may represent effects that, in their own right, serve causally in the pathogenic chain of events. In such a paradigm it is no longer necessary to postulate specific neurotransmitter depletion for specific types of depressive illness. The final common pathway of derangement in the homeostatic mechanisms that normally maintain the reinforcement system in optimal balance may conceivably result from a variety of mechanisms that impair that balance. Janowsky and colleagues[85] for instance, hypothesize that melancholia involves a functional shift in the noradrenergic–cholinergic balance toward the cholinergic side. Stated in neurophysiological terms, depression may result from predominance of the PVS (cholinergic), without any primary change in the MFB (noradrenergic). Similarly, in the thesis of Prange and co-workers[86] changes in the balance between the serotonergic and adrenergic systems rather than the absolute levels of the respective amines become the important consideration. Carroll,[87] extending the balance concept even further, has recently proposed a multiaxial biogenic amine hypothesis involving serotonin, actylcholine, norepinephrine, and dopamine.

The diencephalic final common pathway can be activated in various ways. Situational unipolar depressive states, occurring once in a lifetime, may represent the overwhelming of regulatory function when the individual is confronted with a confluence of environmental factors. On the other hand, recurrent unipolar and bipolar affective states probably require a specific genetic permission that facilitates decompensation with only modest challenge even in the presence of supportive social relationships. A polygenic continuum may explain the immense variety of unipolar–transitional–bipolar phenomena encountered in clinical practice (see Chapter 3). Although the mode of genetic transmission of affective illness has not been conclusively proven, current evidence seems to favor such a model. The consequence of the genetic predisposition may be a functional impairment of the regulatory mechanisms of the biogenic amines such that when stress occurs, early exhaustion or inappropriate balance of the required neurotransmitters results. The possibility that low 5-HIAA, a serotonin metabolite, in the cerebrospinal fluid is a trait marker could emerge as evidence that such a mechanism exists. Similarly the abnormal lithium ratios between red blood cells and plasma in those with bipolar illness may suggest that the encoding for regulation of electrolyte metabolism is disturbed also. Repeated stressors, psychosocial or biological, may then trigger the final common path and induce the new biobehavioral state of melancholia or mania.

This new biobehavioral state is a function of the dynamic behavior of the individual (a complex system) in relation to the environment and not the specific outcome of a single set of inherited variables, however. Therefore as time passes, and the forces upon the individual and within the psychobiological subsystems change, the pathological state usually remits. Ultimately the former mode of adaptive interaction with the environment is assumed once again, although in the natural history of the disorder this may take weeks or months (if death from suicide does not intervene). It is the goal of therapy to speed this natural process. However, while in the new pathological steady state the individual may be very resistant to perturbation, especially by psychosocial means. Thus a relatively stable set of behaviors, sometimes markedly periodic, is characteristic of affective disturbances once they have become established.

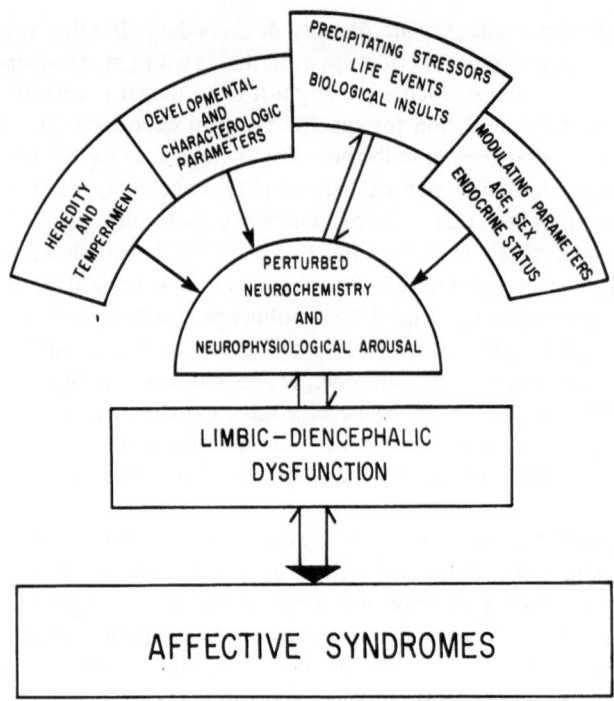

Figure 9.5 Permissive parameters and precipitating variables interact to induce a final common path of diencephalic dysfunction; as this process occurs symptoms of affective illness emerge.

Conclusions

The concept of a final common pathway which is the result of the feedback interaction of diverse events in the diencephalon accounts for the many *shared clinical features* found in a heterogeneous group of depressive disorders, especially once they attain the melancholic phase. On the other hand, the multiple systems that interact, their temporal sequencing, the varied predisposing parameters upon which the precipitating factors impinge; all these elements taken together in a dynamic whole easily account for why *phenomenologic heterogeneity* (e.g., manic and melancholic phases of the illness in the same patient, varied antecedent factors) also exists.

Thus in summary we see melancholia (and mania) as evidence of a new psychobiological state which represents the final common path of various processes, including:

- Life events or chemical stressors that so overwhelm the adaptive skills of those predisposed to affective illness that they see themselves losing control over their destiny
- Escalating levels of subjective turmoil concomitant with psychomotor disturbances
- Heightened neuronal excitability and arousal with interplay of genetically and developmentally vulnerable neuronal circuits in the diencephalon

- Impaired monoaminergic transmission
- Derangement of the neurochemical substrates of reinforcement and their endocrine modulation
- Further breakdown of coping mechanisms
- Vicious cycle of more turmoil, arousal, and hopelessness

The dynamic interplay of these events leads to a loss of the optimum regulation of the subsystems of the diencephalon that integrate and control mood, reward, psychomotor activity, and arousal through both interoceptive and exteroceptive mechanisms. Figure 9.5 outlines these concepts in a simplified manner.

Depression is perhaps the only major reversible psychopathological state that is not confined to human beings. Conceptualized as a reduction or failure of those behaviors that reflect the adaptive interaction of the animal with its surrounding environment, it represents a behavioral state for which there is broad vulnerability. This would explain the occurrence of depressive behaviors in dogs, monkeys, and humans; in children, adults, and the elderly; and in the introverted as well as the extroverted. Although milder forms of "depression" may befall almost anybody, the more severe melancholic depressions seem to involve the interaction of genetic and interpersonal–developmental events; these events impart a special fragility to the diencephalic mechanisms of reinforcement, and to those feedback loops that maintain the centers in homeostasis with contiguous neurophysiological and neuroendocrine systems.

References

1. Bernard, C. *An introduction to the study of experimental medicine.* English edition translated by Greene, H. C. New York: Macmillan, 1927.
2. Vaillant, G. E. *Adaptation to life.* Boston: Little, Brown, 1977.
3. Whybrow, P., and Parlatore, A. Melancholia, a model in madness: A discussion of recent psychobiologic research into depressive illness. *International Journal of Psychiatry in Medicine,* 1973, *4,* 351–378.
4. Akiskal, H. S., and McKinney, W. T., Jr. Depressive disorders: Toward a unified hypothesis. *Science,* 1973, *182,* 20–29.
5. Akiskal, H. S., and McKinney, W. T., Jr. Overview of recent research in depression: Integration of ten conceptual models into a comprehensive clinical frame. *Archives of General Psychiatry,* 1975, *32,* 285–305.
6. Hendrie, H. C. Organic brain disorders: Classification, the "symptomatic" psychoses, misdiagnosis. *Psychiatric Clinics of North America,* 1978, *1,* 3–19.
7. Whybrow, P. C., and Hurwitz, T. Psychological disturbances associated with endocrine disease and hormone therapy. In E. J. Sachar (Ed.), *Hormones, behavior and psychopathology.* New York: Raven Press, 1976, pp. 125–143.
8. Krauthammer, C., and Klerman, G. L. Secondary mania: Manic syndromes associated with antecedent physical illness or drugs. *Archives of General Psychiatry,* 1978, *35,* 1333–1339.
9. Akiskal, H. S. The bipolar spectrum: New concepts in classification and diagnosis. In L. Grinspoon (Ed.), *Psychiatry update: The American Psychiatry Association annual review, vol. II.* Washington, D.C.: American Psychiatric Press, 1983, pp. 271–292.
10. Akiskal, H. S., Rosenthal, R. H., Rosenthal, T. L., Kashgarian, M., Khani, M. K., and Puzantian, V. R. Differentiation of primary affective illness from situational, somatic and secondary depressions. *Archives of General Psychiatry,* 1979, *35,* 635–643.
11. Kupfer, D. J., and Thase, M. E. The use of the sleep laboratory in the diagnosis of affective disorders. *Psychiatric Clinics of North America,* 1983, *6,* 3–25.
12. Whybrow, P. C., and Mendels, J. Toward a biology of depression: Some suggestions from neurophysiology. *American Journal of Psychiatry,* 1969, *125,* 1491–1500.

13. Greden, J. F., and Carroll, B. J. Psychomotor function in affective disorders: An overview of new monitoring techniques. *American Journal of Psychiatry*, 1981, *138*, 1441–1448.
14. Widlöcher, D. J. Psychomotor retardation: Clinical, theoretical, and psychometric aspects. *Psychiatric Clinics of North America*, 1983, *6*, 27–40.
15. Knotting, L. von, Perris, C., and Strandman, E. Diurnal variations in intensity of symptoms in patients of different diagnostic groups. *Archiv für Psychiatrie und Nervenkrankheiten*, 1977, *224*, 295–312.
16. Klein, D. Endogenomorphic depression: A conceptual and terminological revision. *Archives of General Psychiatry*, 1974, *31*, 447–454.
17. Kraines, S. H. *Mental depressions and their treatment*. New York: Macmillan, 1957.
18. Schildkraut, J, J., Osulak, P. J., Shatzberg, A. F., and Rosenbaum, A. H. Relationship between psychiatric diagnostic groups of depressive disorders and MHPG. In J. W. Maas (Ed.), *MHPG: Basic mechanisms and psychopathology*. New York: Academic Press, 1983, pp. 129–144.
19. Van Praag, H. M. The significance of biological factors in the diagnosis of depression. I. Biochemical variables. *Comprehensive Psychiatry*, 1982, *23*, 124–135.
20. Carroll, B. J., Feinberg, M., Greden, J. F., Tarika, J., Albala, A. A., Haskett, R. F., James, N. M., Kronfol, Z., Lohr, N., Steiner, M., Vigne, J. P., and Young, E. A specific laboratory test for the diagnosis of melancholia. *Archives of General Psychiatry*, 1981, *38*, 15–22.
21. Loosen, P. T., and Prange, A. J. Serum thyrotropin response to thyrotropin-releasing hormone in psychiatric patients: A review. *American Journal of Psychiatry*, 1982, *139*, 405–416.
22. Winfree, A. T. *The geometry of biological time*. New York: Springer-Verlag, 1980.
23. Winokur, G., and Clayton, P. Family history studies. I. Two types of affective disorder separated according to genetic and clinical factors. In J. Wortis (Ed.), *Recent advances in biological psychiatry*, New York: Plenum Press, 1967.
24. Perris, C. The distinction between bipolar and unipolar affective disorders. In E. S. Paykel (Ed.), *Handbook of affective disorders*. New York: Guilford Press, 1982, pp. 45–58.
25. Mendelewics, J., and Rainer, J. D. Adoption study supporting genetic transmission in manic depressive illness. *Lancet*, 1977, *268*, 327–329.
26. Cadoret, R. J. Psychopathology in adopted-away offspring of biologic parents with antisocial behavior. *Archives of General Psychiatry*, 1978, *35*, 176–184.
27. Gershon, E. S., Dunner, D. L., and Goodwin, F. K. Toward a biology of affective disorders. *Archives of General Psychiatry*, 1971, *25*, 1–15.
28. Fieve, R. Overview of therapeutic and prophylactic trials with lithium in psychiatric patients. In S. Gershon and B. Shopsin (Eds.), *Lithium: Its role in psychiatric research and treatment*. New York: Plenum Press, 1973, pp. 317–349.
29. Sedvall, G. Neurotransmitter disturbances and predisposition to depressive illness. *Advances in Biological Psychiatry*, 1981, *7*, 26–33.
30. Dorus, E., Pandey, G. N., Frazer, A., and Mendels, J. Genetic determinant of lithium ion distribution: An in vitro monozygotic–dizygotic twin study. *Archives of General Psychiatry*, 1974, *31*, 463–465.
31. Gershon, E. S. Overview of genetic trait markers in affective illness. Paper presented at the 21st meeting of the American College of Neuropsychopharmacology, San Juan, Puerto Rico, 1982.
32. Sitaram, N., Nurnberger, J. I., Gershon, E. S., and Gillin, C. Cholinergic regulation of mood and REM sleep: Potential model and marker of vulnerability to affective disorder. *American Journal of Psychiatry*, 1982, *139*, 571–576.
33. Kraepelin, E. *Manic–depressive insanity and paranoia*. Edinburgh; E and S Livingstone, 1921.
34. Akiskal, H. S. Dysthymic and cyclothymic disorders: A paradigm for high-risk research in psychiatry. In J. M. Davis and J. W. Maas (Eds.). *The affective disorders*. Washington, D.C.: American Psychiatry Press, 1983, pp. 211–231.
35. Akiskal, H. S. Characterologic manifestations of affective disorders: Toward a new conceptualization. *Integrative Psychiatry*, 1984 (in press).
36. Angst, J., Felder, W., and Frey, R. The course of unipolar and bipolar affective disorders. In M. Schou and E. Strömgren (Eds.). *Origin, prevention and treatment of affective disorders*. London: Academic Press, 1979, pp. 215–239.

37. Weissman, M. M., and Klerman, G. L. Sex differences in the epidemiology of depression. *Archives of General Psychiatry*, 1977, *34*, 98–111.

38. Robinson, D. S., Davis, J. M., Nies, A., Ravaris, C. L., and Sylvester, D. Relation of sex and aging to monoamine oxidase activity of human brain, plasma, and platelets. *Archives of General Psychiatry*, 1971, *24*, 536–539.

39. Prange, A. J., Wilson, J. C., Wabon, A. M., and Lipton, M. A. Enhancement of imipramine antidepressant activity by thyroid hormone. *American Journal of Psychiatry*, 1969, *126*, 457–469.

40. Coppen, A., Whybrow, P. C., Noguera, R., Maggs, R., and Prange, A. J. The comparative antidepressant value of 1-tryptophan and imipramine with and without attempted potentiation by liothyronine. *Archives of General Psychiatry*, 1972, *26*, 234–241.

41. Dalton, K. *The premenstrual syndrome*. Springfield, Illinois: C. C. Thomas, 1964.

42. Brockington, I. F., and Kumar, R. (Eds.). *Motherhood and mental illness*. London: Academic Press, 1982.

43. Neugarten, B. L., and Kraine, R. J. Menopausal symptoms in women of various ages. In L. Gitman (Ed.) *Endocrines and aging*. Springfield, Illinois: C. C. Thomas, 1967, pp. 218–230.

44. Kane, F. J. Iatrogenic depression in women. In W. E. Farn, I. Karacan, A. D. Pokorny, and R. L. Williams (Eds.), *Phenomenology and treatment of depression*. New York: Spectrum Publications, 1977, pp. 69–80.

45. Goodwin, F., and Bunney, W. E., Jr. Depression following reserpine: A re-evaluation. *Seminars in Psychiatry*, 1971, *3*, 435–448.

46. Mendels, J., and Frazer, A. Brain biogenic amine depletion and mood. *Archives of General Psychiatry*, 1974, *30*, 447–451.

47. Kraemer, G. W., and McKinney, W. T. Interaction of pharmacologic agents which alter biogenic amine metabolism in depression. *Journal of Affective Disorders*, 1979, *1*, 33–54.

48. Akiskal, H. S. Factors associated with incomplete recovery in primary depressive illness. *Journal of Clinical Psychiatry*, 1982, *43*, 266–271.

49. Schmale, A. H., and Engel, G. L. The role of conservation-withdrawal in depressive reactions. In E. J. Anthony and T. Benedek (Eds.), *Depression and human existence*. Boston: Little, Brown, 1975, pp. 183–198.

50. Bowlby, J. The making and breaking of affectional bonds. I. Aetiology and psychopathology in the light of attachment theory. *British Journal of Psychiatry*, 1977, *130*, 201–210.

51. Perris, C. A. A study of bipolar (manic–depressive) and unipolar recurrent depressive psychoses. *Acta Psychiatrica Scandinavica*, 1966, *42*,(Suppl.), 7–188.

52. Beck, A., Sethi, B., and Tuthill, R. Child bereavement and adult depression. *Archives of General Psychiatry*, 1963, *9*, 295–302.

53. Levi, D. L., Fales, C. L., Stein, M., and Sharp, V. H., Separation and attempted suicide. *Archives of General Psychiatry*, 1966, *15*, 158–164.

54. Hill, O. The association of childhood bereavement with suicide attempt in depressive illness. *British Journal of Psychiatry*, 1969, *115*, 301–334.

55. Akiskal, H. S., Bitar, A. H., Puzantian, V. R., Rosenthal, T. L., and Walker, P. W. The nosological status of neurotic depression: A prospective three-to-four-year examination in light of the primary–secondary and unipolar–bipolar dichotomies. *Archives of General Psychiatry,* 1978, *35*, 756–766.

56. Rosenthal, T. L., Akiskal, H. S., Scott-Strauss, A., Rosenthal, R. H., and David, M. Familial and developmental factors in characterological depressions. *Journal of Affective Disorders*, 1981, *3*, 183–192.

57. Chodoff, P. The depressive personality: A critical review. *Archives of General Psychiatry*, 1972, *27*, 666–673.

58. Akiskal, H. S., Hirschfeld, R. M. A., and Yerevanian, B. I. The relationship of personality to affective disorders: A critical review. *Archives of General Psychiatry*, 1983, *40*, 801–810.

59. Whybrow, P. C. Adaptive styles in the etiology of depression (Lesson 29). In F. Flach (Ed.), *Directions in psychiatry*. New York: Hatherleigh, 1981.

60. Von Zerssen, D. Premorbid personality and affective psychoses. In G. D. Burrows (Ed.), *Handbook of studies on depression*. Amsterdam: Excerpta Medica, 1977.

61. Cadoret, R. J., Baker, M., and Dorzab, J. Depressive disease: Personality factors in patients and their relatives. *Biological Psychiatry*, 1971, *3*, 85–93.

62. Hirschfeld, R. M., and Klerman, G. L. Personality attributes and affective disorders. *American Journal of Psychiatry*, 1979, *136*, 67–70.

63. Hirschfeld, R. M. A., Klerman, G. L., Clayton, P. J., and Keller, M. Personality and depression: Empirical findings. *Archives of General Psychiatry*, 1983, *40*, 993–998.

64. Hirschfeld, R. M. A., Klerman, G. L., Clayton, P. J., and Keller, M. Assessing personality: Effects of the depressive state on trait measurement. *American Journal of Psychiatry*, 1983, *140*, 695–699.

65. Metcalfe, M. The personality of depressive patients. In A. Coppen (Ed.), *Recent developments in affective disorders*. Ashford: Headley Brothers, 1968.

66. Clayton, P. J. The sequelae and nonsequelae of conjugal bereavement. *American Journal of Psychiatry*, 1979, *136*, 1530–1534.

67. Spitz, R. Anaclitic depression: An inquiry into the genesis of psychiatric conditions in early childhood. *Psychoanalytic Study of the Child*, 1942, *2*, 313–342.

68. Blazer, D. G. Impact of late-life depression on the social network. *American Journal of Psychiatry*, 1983, *140*, 162–166.

69. Flaherty, J. S., Gaviria, F. M., Black, E. M., Altman, E., and Mitchell, T. The role of social support in the functioning of patients with unipolar depression. *American Journal of Psychiatry*, 1983, *140*, 473–476.

70. Paykel, E. S., Myers, J. K., Dienelt, M. N., Klerman, G. L., Lindenthal, J. J., and Pepper, M. P. Life events and depression: A controlled study. *Archives of General Psychiatry*, 1969, *21*, 753–760.

71. Fava, G. A., Munari, F., Pavan, L., and Kellner, R. Life events and depression: A replication. *Journal of Affective Disorders*, 1981, *3*, 159–165.

72. Ambelas, A. Psychologically stressful events in the precipitation of manic episodes. *British Journal of Psychiatry*, 1979, *135*, 15–21.

73. Racamier, P. C., and Blanchard, M. De l'angiosse à la manie. *L'Évolution Psychiatrique*, 1957, *3*, 558–587.

74. Bunney, W. E. Psychopharmacology of the switch process in affective illness. In M. A. Lipton, A. Dimasco, and K. F. Killam (Eds.), *Psychopharmacology: A generation of progress*. New York: Raven Press, 1978, pp. 1249–1259.

75. Gerner, R. H., Post, R. M., and Bunney, W. E. A dopaminergic mechanism in mania. *American Journal of Psychiatry*, 1976, *133*, 1177–1180.

76. Frankenhaeuser, M. Experimental approaches to the study of human behaviour as related to neuroendocrine functions. In L. Levi (Ed.), *Society, stress and disease*. London: Oxford University Press, 1971.

77. Mason, J. W. Organisation of the multiple endocrine responses to avoidance in the monkey. *Psychosomatic Medicine*, 1968, *30*, 774–790.

78. Barchas, J., and Freedman, D. Brain amines: Response to physiological stress. *Biochemical Pharmacology*, 1963, *12*, 1232–1235.

79. Bliss, E., and Zwanziger, J. Brain amines and emotional stress. *Journal of Psychiatric Research*, 1966, *4*, 189–198.

80. Weiss, J. M., Glozer, H. I., Pohorecky, L. A., Bailey, W. H., and Schneider, L. H. Coping behavior and stress-induced behavioral depression: Studies of the role of catecholamines. In R. A. Depue (Ed.), *The psychobiology of the depressive disorders*. New York: Academic Press, 1979.

81. Anisman, H., and Zacharko, R. K. The predisposing influences of stress. *Behavioural and Brain Sciences*, 1982, *5*, 89–137.

82. Sweeney, D. R., Maas, J. W., and Heninger, G. R. State anxiety, physical activity and urinary 3-methoxy-4-hydroxy-phenethylene glycol excretion. *Archives of General Psychiatry*, 1978, *35*, 1418–1423.

83. Stein, L. Chemistry of reward and punishment. In D. Efron (Ed.), *Psychopharmacology: A review of progress, 1957–1967*. Washington, D.C.: U.S. Government Printing Office, 1968, pp. 105–123.

84. Crow, T. Catecholamine-containing neurones and electrical self-stimulation. II. A theoretical interpretation and some psychiatric implications. *Psychological Medicine*, 1973, *3*, 66–73.

85. Janowsky, D., El-Yousef, K., Davis, M., and Sekerke, H. J., A cholinergic–adrenergic hypothesis of mania and depression. *Lancet*, 1972, *2*, 632–635.

86. Prange, A. J., Wilson, I. C., and Lynn, C. W. L-tryptophan in mania: Contribution to the permissive hypothesis of affective disorders. *Archives of General Psychiatry*, 1974, *30*, 56–62.
87. Carroll, B. J. Problems of paradox, definition and method in depression. Paper presented at the Dalheim Conference on the Origins of Depression, Berlin, November, 1982.

Implications for Clinical Practice, Training, and Research

The plan we have outlined (psychobiology) offers a satisfactory place for the upshot of practically every other scheme of pathology and therapy. It is nondogmatic, but critically discriminating; although it frankly places itself on the ground of the best common experience and the best trained common sense rather than the authority derived from scientific system on ground of limiting theory.

Adolf Meyer[1]
The Problem of Mental Health (1934)

Introduction

In this book we have tried to present a pluralistic and integrative view of the psychobiology of mood disorders. Our goal has been to present a holistic paradigm for understanding mood disorders, but the approach can serve equally well for understanding other major psychiatric syndromes. Despite the significant advances made over the last two decades, we are only just beginning to understand the affective psychoses. Undoubtedly, our present limited understanding of the monoaminergic and endocrine mechanisms that underlie the various types of depression will give way to more sophisticated theories. Likewise, the mechanisms by which attachment in early development influences mood and depression later in life will become clearer. Most importantly, the mechanisms whereby psychologically defined events such as developmental deprivation, specific characterological traits, and adult loss *interact* with predisposing heredity and concurrent physiological stressors will be elucidated.

The way in which these many questions are approached in clinical practice, training, and research will have important implications for the speed and depth at which we gather further knowledge. We believe that no *single* viewpoint can provide a sufficient explanation for depression or mania. These disorders must be understood in a multidimensional framework as illnesses representing the common dysfunctional pathway which results from the interaction of a diverse range of influences—genetic, familial, developmental, interpersonal, and neurobiological. In this final chapter we review such an approach from the practical standpoint.

Implications for Clinical Practice

The Importance of Diagnosis

Given our present understanding, it is no longer acceptable to evaluate and treat a heterogeneous group of illnesses such as mood disorders within the theoretical framework of one approach—be that analytic, cognitive, behavioral, or biological.

The clinician treating affective illness today should be well grounded in descriptive psychopathology and nosology. The answers to specific questions must be sought. It is necessary to determine whether the complaint of depression is a transient mood occasioned by situational misfortune, or a transient symptomatic phase of a nonaffective disorder such as alcoholism, schizophrenia, or somatization disorder. If primary and sustained, has the depression reached syndromal status? Have there been previous episodes, either depressive or manic? Is the depression chronic; has it persisted for more than 2 years? Is it superimposed upon a temperamental disorder such as cyclothymia? Has the depression acquired melancholic quality, with autonomous course and significant disturbance in psychomotor, vegetative, and pleasure mechanisms (the "endogenous" clinical picture)? If there are psychotic symptoms, are they mood-congruous? Careful consideration of these questions should make it possible to classify the patient's illness into one of the clinical subtypes of affective disorder.

Some clinicians object to the assignment of formal descriptive diagnoses, arguing that this approach is dehumanizing and irrelevant to treatment. The current evidence indicates serious flaws in such arguments. While the uniqueness of each patient must be respected, it is what the patient *shares* with other patients that is most relevant to treatment. These shared characteristics make up the diagnostic rubric. The clinician who disregards diagnostic considerations will be unable to bring to the patient the benefit of past clinical and research wisdom gained from similarly diagnosed cases. In effect, reluctance to make use of modern descriptive nosology may lead to inadequate clinical care.[2] Fortunately, the recent formulation of explicit operational criteria for psychiatric diagnosis, as exemplified by *DSM-III*,[3] has now provided a reasonably reliable method for the practical categorization of psychopathological syndromes.

Principles of Comprehensive Clinical Care

In addition to the chronology of symptoms, one must take *a thorough genetic–family history*. In earlier chapters we have reviewed the increased risk for depression in individuals with an established family history of affective illness. One does not have to be a genetic determinist to appreciate the importance of this information. Family history aids subclassification of an affective state—especially with respect to the unipolar–bipolar dichotomy. As discussed in Chapter 3 the presence of bipolar familial history in a unipolar patient tends to decrease the patient's threshold for hypomanic episodes—whether spontaneous or pharmacologically induced—and predicts an illness with high episode frequency compared with other depressions. The implications of family history for alcoholism, on the other hand, are quite different, for they predict serious characterological disturbances and a tempestuous course. One also needs to know how other family members with the illness have been treated, since responsiveness to drugs often runs in families also. Further, the family situation must be explored; negative attitudes toward depression and mental illness in a family with affected members can adversely influence the course of the affective episode in the identified patient.

There are several reasons why *a detailed developmental history* is also crucial for treatment decisions. Such traits as negative self-esteem and passive-dependence are frequent and important in the clinical history of depressed patients. A careful history from the patient and significant people in the patient's immediate environment both during and after the depression may help to clarify whether these traits are characterological or secondary to the depression. If they are of developmental origin, long-term psychotherapeutic intervention may be indicated. When such traits are secondary to the depressive episode and are not prominent during illness-free periods, some kind of supportive work may be all that is required to enhance the therapeutic effect of antidepressant medication. Developmental object loss is also relevant to adult depressive illness, because specific losses at certain ages may sensitize the individual to a higher risk of certain complications such as characterological difficulties and suicide attempts.

In addition to genetic–familial and early developmental history, *current role conflicts and adult developmental crises* are of clinical relevance. It has been recently documented that psychotherapy aimed at such practical issues interacts positively with pharmacotherapy. Although the etiological role of adult life events is usually one of precipitation, sufficiently devastating experiences may even have causal influence. One must be also attuned to the possibility that life crises and role conflicts can be epiphenomenal; for instance, irritable and morose people in the syndromal phase of depressive illness may alienate and lose loved ones. When the affective basis of such conflicts is not recognized—and proper treatment for the primary disorder is not instituted—familial and socioeconomic disasters may ensue.

The clinician must be particularly alert to the fairly common situation in which *intermittent subsyndromal affective disorder is masked by an interpersonal, conjugal, or behavioral problem.* For instance, research conducted at the University of Tennessee[4] has shown that cyclothymic individuals often do not complain of mood swings; repeated marital failure or romantic break-ups, episodic promiscuity, alcohol and drug abuse, uneven work and school record, geographic instability, and dilettantism are the prominent clinical manifestations. Extensive clinical evaluation and long-term follow-up are often necessary to support the diagnosis of a cyclothymic bipolar illness in these patients. These patientsare often exposed to individual and group psychotherapy and a host of nonspecific (i.e., other than lithium) chemotherapies with little appreciable benefit.

Another important aspect of the caring for the affectively ill is the need *to be flexible in one's approach over time.* Depression and mania are often recurrent and, as is true with any lifelong illness, a patient's needs change. The clinician who works with affectively ill persons should therefore avoid prescribing one form of treatment throughout the course of the illness. One hopes that we have passed the era when patients were accepted for treatment depending on whether they were good candidates for a given type of therapy. Thus, the clinical care of the affectively ill envisions the proper balance of various chemotherapies and psychotherapies in a flexible, individualized treatment plan that takes into account the course of the illness. As an example, tricyclics and supportive psychotherapy in an outpatient setting over 6 to 9 months may be sufficient treatment for a first depressive episode in an individual with no marked character pathology. Should mania be precipitated by tricyclic treatment, hospitalization may be required to avert social, marital, or financial crises. For a first manic attack or in bipolar patients with *infrequent* episodes, lithium therapy is usually instituted for 6 months at a time, and the patient and significant others are educated about the recurrent nature of the illness. If there are recurrences more often than every 2 or 3 years, lithium prophylaxis is appropriate. At this stage, the impact of coming to grips with a new identity that accommodates the concept of a recurrent psychiatric illness becomes an important issue in psychotherapy. Compliance with lithium treatment may become a problem in the patient whose self-esteem depends on hypomanic performance or whose mania has permitted the acting out of certain impulses toward a parent, boss, or spouse.[5] A behavioral approach to compliance may sometimes succeed in such cases. Educative sessions with family, spouse, or other significant people are usually mandatory. Another way to foster compliance is to hold group therapy sessions of a number of bipolar patients with a balanced mixture of compliant and noncompliant patients.

Combining psychotherapy and pharmacotherapy is both feasible and desirable when treating the affectively ill. Studies by Klerman, Weissman, and associates have shown that these two forms of treatment influence different processes.[6,7] Tricyclics are particularly effective in alleviating symptoms that reflect biological dysfunction; psychotherapy seems to have its most salutary effect on inhibited communication and role conflicts. As symptoms of the illness gradually recede, the patient becomes more able to cope and to benefit from the enhanced social functioning that psychotherapy can bring.

The timing and sequence of therapies is also crucial, especially in melancholia (autonomous depressions). Such patients are too ill—affectively hyperaroused or labile, unable to concentrate and preoccupied with negative cognitions—to benefit from psychotherapy, and predictably get worse when psychological therapies are given alone.[8] In fact, systematic exploration of the personal problems of an acutely depressed individual may augment guilt and hopelessness and lead to suicidal behavior. Reversal of the symptom picture with somatic therapy is ordinarily required for the patient to become accessible to any elaborate psychotherapeutic intervention. Although this principle is usually applied to manic and schizophrenic patients, it is sometimes neglected in treating clinical depression.

Mood Clinics and Affective Disorder Units

Recently there has been an emergence of specialized affective disorder clinics, or mood clinics, that address many of the diagnostic and therapeutic questions raised in the foregoing discussion. Psychiatry has lagged behind the rest of medicine in developing such subspecialty clinics centered around a disease entity or an organ system. Instead, treatment has been organized around therapeutic modalities, such as psychodynamic therapy, behavior therapy, chemotherapy, group therapy, and biofeedback. Acceptance of patients into such treatments has often depended on their appropriateness for the modality in question. By contrast, mood clinics, like other emerging subspecialty clinics (e.g., pain, sexual dysfunction, and phobia clinics), are organized around specific syndromes or groups of related disorders. Criteria for whom to accept into the clinic are based on the illness rather than criteria for good outcome or a given mode of therapy.

Most of these new syndrome-oriented clinics have been set up in the outpatient setting. In mood clinics the emphasis is on keeping patients with recurrent affective disorders out of the hospital and on enhancing their symptomatic, interpersonal, and vocational recovery. Such clinics are in the best tradition of community psychiatry. Thus, the proven efficacy and safety of lithium salts and tricyclic antidepressants in preventing recurrences of mood disorders have provided relative freedom from socially disruptive and disabling episodes, thereby permitting the deployment of psychotherapeutic and sociotherapeutic efforts on a long-term basis.

The new and firmer evidence than was available in the past about biological contributions in the causation of mood disorders has legitimized the testing and use of sophisticated and powerful biological technologies in these new clinics. However, the use of lithium and tricyclic drugs mandates a solid grounding in pharmacology, internal medicine, and neurology in those responsible for the clinics' operation.[9]

Although combined biological and psychosocial interventions on an outpatient basis are ordinarily successful in preventing affective episodes, an inpatient stay is sometimes necessary for proper diagnostic work-up, emergency management, and the treatment of serious relapse. Psychiatric inpatient wards to meet these needs must embrace the practical implications of recent advances in our understanding of affective illness. The "average" unit typically first "excludes" medical disease and then proceeds to treat depression as a "functional" disorder. We propose that psychiatry wards that admit a large number of affectively ill patients should be equipped to undertake such diagnostic laboratory investigations as sleep encephalogram (EEG), dexamethasone suppression, thyroid studies, plasma levels of psychoactive drugs, and urinary collection of biogenic amine metabolites as aids in differential diagnosis and in predicting drug responsiveness. This list is illustrative rather than comprehensive and will obviously change with new research developments. These tests offer a beginning approach to ordering the biological concomitants of affective illness. Although it is premature to predict their absolute value as diagnostic adjuncts and monitoring tools,[10] several examples of their range of practical usefulness can be given:

1. The dexamethasone suppression test (DST), which has fair specificity for melancholia (provided appropriate medical and drug exclusions are made),[11]

can be used to *clarify* and *support* an affective diagnosis, particularly when neurotic and characterological disorders appear prominently in the list of differential diagnoses.[10,12] Furthermore, the persistence of a DST abnormality after apparent clinical recovery may indicate early relapse and suggests the value of continued treatment.

2. Blunted thyroid-stimulating hormone (TSH) response to thyrotropin-releasing factor (TRH)[13] may help in differentiating mania from schizophrenia.[14] Furthermore, an exaggerated response of TSH to TRH suggests thyroid deficiency and the need for replacement hormone therapy in a percentage of unipolar women.[15]

3. Short rapid-eye-movement (REM) latency on the sleep EEG,[16] especially on two consecutive nights,[17] is usually associated with primary as opposed to secondary depressions (when drug withdrawal, narcolepsy, and certain rare sleep disorders are excluded). Rapid-eye movement density, the amount of eye movements per unit of REM period, is often decreased in depressions occurring in the setting of diffuse brain disease or serious systemic disease with brain involvement.[18]

4. Brief pharmacological switches to hypomania upon tricyclic drug administration in a depressed patient (with no previous elated periods) predict the future course of spontaneous hypomania.[19,20] Likewise, lowered urinary excretion of MHPG appears to be a correlate of bipolar depression.[21]

5. The rate of tricyclic response may be increased by 10%–15% with appropriate dosage adjustment as revealed through plasma tricyclic determinations.[22]

Ideally, the administration and interpretation of such methods requires a specialized affective disorder unit or subunit of a general service ward. Just as in general medicine services have slowly specialized, so do we expect such specialization to occur in psychiatry. Indeed, we believe that the present knowledge base in psychiatry, though ever-expanding and changing, already justifies a reorganization of our inpatient psychiatric health care delivery system along syndromal lines.

A related need is for the development of in- and outpatient services specifically for mood disorders on a regional or state level. The diagnosis and treatment of affective disorders has now become sufficiently complicated that many smaller communities are unable to address patients' needs adequately.[23] Mental health planners have tended not to recognize this, and local communities have been pressured to provide a comprehensive range of services. The generalist, whether in psychiatry or primary care, can treat the "usual" case of depression. However, recurrent affective disorders present many clinical challenges that the generalist may be too inexperienced to handle. The existence of a series of back-up clinics and inpatient units specialized in the diagnosis and treatment of mood disorders will enable referral for these more complicated cases. Although the optimal organization of mental health care delivery systems will vary depending on the disorder being treated, it is our view that programs for affective disorders would benefit from the format outlined here.

Finally, we believe it is important that all patients suspected of having an affective disorder be evaluated by a psychiatrist knowledgeable and skilled in the area. This does not include everyone who has the "blues," but rather those individuals with *inca-*

pacitating, pervasive, or *frequent* mood swings.[24] The rapid advance of specialized knowledge regarding affective disorders makes such specialized psychiatric referral important.

Implications for Training

In General

What are the implications of the integrated psychobiological approach for training? The first and most important one is that *all* mental health professionals should receive substantial training in the area of affective disorders. This training should transcend ideologic preferences and provide exposure to a multidisciplinary, multivariate view of depression. The primary goal is to achieve an understanding of affective disorders as a group of illnesses—not just alternative ways of life—that have their own unique symptomatology, morbidity, course, and potential mortality.

The development of specialized clinics and inpatient services will foster better training in the broad area of affective disorder. Such services can provide trainees with an intensive and longitudinal exposure to a diversified cohort of affectively ill patients, combined with systematic readings and didactic clinical instruction. It is critical that primary care physicians and all mental health professionals have the experience of working in these or similar settings.

The opportunity to diagnose and follow longitudinally a large group of patients representing all clinical subtypes of affective illness is particularly important for the psychiatric resident. The episodic nature of major affective disorders makes longitudinal exposure essential. In actual practice one will of necessity follow the affective swings of such patients over the course of decades.

Post-Residency Fellowships

The development of post-residency fellowships in mood disorders should be considered a high priority in psychiatric education. Psychiatry today is developing along three broad avenues: (1) an emphasis on data gathering, (2) closer identification with general medicine, and (3) integration of the biological and psychosocial approaches to mental disorders. Such developments are especially visible in the area of affective illness, which has attracted major clinical and research interest during the past decade. The growth of affective disorder clinics provides a fertile ground for training future psychiatric specialists in systematic clinical observation, research methodology, medical psychiatry, and a theoretical framework that derives from multidisciplinary sources. The emphasis is on subspecialization in affective disorders rather than specialization in one discipline or school of psychiatric thinking. The clinical setting is analogous to medical subspecialities which focus on one class of disorders and utilize all available approaches in the understanding and treatment of disorders within that subspecialty. This is in contrast to many traditional training settings in psychiatry which emphasize a particular school of thought, e.g., psychoanalysis, behaviorism, and biological psychiatry. Although we do not deny the important contributions that these different schools have brought to psychiatry, the future of psychiatry appears intimately linked with specialization along syndromal lines.

An affective disorder clinic provides an excellent opportunity to integrate clinical service and academic pursuits: the quality of care is enhanced by trainees with special knowledge and training, and systematically recorded data is vital for clinical research. The trainee undertakes the diagnosis and treatment of the depressive illness in addition to the responsible care of the person suffering from it. He or she is taught a systematic approach to psychiatric evaluation and diagnosis that is validated by the trainees' own longitudinal observation. Trainees are expected to become knowledgeable in the integrated and differential use of drugs and verbal therapies, based upon a reasoned understanding of their interaction. In summary, an affective disorder clinic can particularly enhance the conceptual training of future psychiatrists by (1) emphasizing a multifactorial model for the pathogenesis of affective disorders, (2) requiring critical appraisal of the literature, (3) fostering respect for data and phenomenology, (4) subjecting interpersonal and intrapsychic variables to the same kind of rigorous hypothesis-testing as biological variables, and (5) presenting the complexity, uncertainty, and diversity in our field as an intellectual challenge that makes psychiatry uniquely exciting.

Implications for Research

What are the implications of an integrated perspective for future research? Having summarized the specific areas of ongoing inquiry in earlier chapters, we will confine ourselves to general remarks.

In research it is first necessary to isolate specific and measurable factors, to develop the appropriate methodology to evaluate and measure them, and then, after coming to some understanding of the parts, attempt an heuristic synthesis. The two activities, reduction and synthesis, are mutually interdependent; inevitably, in seeking validation of the behavior of the whole, they lead to studies of interaction between the parts. Ideally, studies in affective illness must now enter this most complex phase of investigation. It is obviously difficult to study the interaction between different kinds of variables in systems that demand different techniques for gathering information, but this is the next challenge before us. We can no longer hope that depression is caused by a single disturbance in neurotransmitter mechanisms; nor does object loss specifically lead to clinical depression. It is the interaction, over time, among genetic, developmental, intrapsychic, sociological, and neurobiological variables that becomes the crucial focus.

Longitudinal clinical assessment must remain a cornerstone of our future effort. Formal specialty clinics will aid in this activity, but without a serious commitment to careful data-gathering prospectively over extended time periods, the true value of new information from pharmacology or biology will not be adequately assessed. There is also still much to learn about the syndromes that lie at the periphery of melancholia and mania; it is here that the subtle intertwining of biological and psychosocial variables is frequently reflected in chronically marginal adaptation.[25] Without the perspective of time, such individuals are frequently dismissed and little interest is expressed in them by clinician or researcher.

Formal prospective studies are ideal for study of the interdependence of biological and social factors, but are very difficult to finance and administer over an extended time frame. Even if the children of persons suffering affective disorder are selected for follow-up, the chance of unipolar illness being expressed is only about 15%–20%, scarcely higher than the general population. Furthermore, one must wait 20–40 years for the results. Thus such studies are only rarely achieved. The Grant study[26] (see Chapter 9) is an exception. Careful family studies which review the incidence of illness in two or three generations can enhance the yield to be expected in a prospective cohort, however, especially if combined with studies of *genetic markers*. As more of these state-independent variables are identified in depressed and manic persons, this promises to be a very fruitful area of research.

A better understanding of the syndromes that may reflect *depressive disturbance in children at risk* would also be helpful. This would allow a closer investigation of the interaction of psychosocial stressors and biological response as is classically exemplified by Engel and Schmale's study of Monica.[27] Here again the informed and curious clinician is essential to the development of the more formal research effort; only those following families with affective illness are in a suitable position to conduct the preliminary investigation of potentially pathological syndromes in "normal" children.

Animal models of affective illness, particularly in primates, can continue to be of help to us. Such models have been widely used in other areas of medicine to study both the cause and treatment of disease. The special value of an animal model in relation to depression lies in the study of the multitude of factors that contribute to a phenomenon. With animal models we can manipulate biological and social factors in ways not often practical or ethical in the human being. In primates such models have made it possible to study the interaction of drugs that are known to alter amine metabolism with psychopathology socially induced by means of separation experiences. Altering the metabolism of certain biogenic amines does indeed potentiate the despair response of young monkeys to separation, for example, while manipulation of other amine systems has no effect or actually counteracts the despair response. Other work has highlighted the complexity of the relationships among brain biogenic amines in the maintenance of normal behavior. By certain experimental techniques, it is possible to deplete by 99% the brain norepinephrine of a rhesus monkey. Such depletion produces acute behavioral changes, including social withdrawal and decreased activity. However, these behavioral changes do not persist even in the face of continued norepinephrine depletion, emphasizing again the adaptive resilience of the normal central nervous system and the limitations of single-variable hypotheses regarding affective illness.[28]

Data from such models can thus provide useful insights into the interactional mechanisms behind human disorder. The most valuable assets they offer are undoubtedly the control of the social environment and the opportunity to employ powerful pharmacological probes in an overlapping paradigm. They do, however, have their limitations; it is not easy to obtain biological materials for assessment (sometimes more difficult than in human beings!), and sacrifice to study brain amines directly is rarely a profitable enterprise. Primate colonies are difficult to start and expensive to maintain, and the considerable time invested in development of the "model" animals is not easily rewarded by the temporally isolated fragments of information that are obtained by destroying them. Hence *basic studies in biochemical pharmacology* in the laboratory rat are unlikely to be superseded. It is these studies that will continue to provide the fundamental building blocks for our understanding of the basic mechanisms of neuronal function and adaptation.

In the clinical arena the use of *neuroendocrine measures,* especially challenge tests, will become of increasing value as indices of central monoamine function. These procedures can potentially serve as a bridge between the data obtained from small animals and the disturbed biological state of the illness. The DST and the TRH test were early examples of challenge paradigms which have been used as general probes of hypothalamic arousal. Growth hormone response, prolactin levels, the circadian rhythm of melatonin; these are also emerging as useful tools. As we come to understand more precisely the neurotransmitter drive systems that govern the release of these hormones, their utility as indices of the balance between amine systems in the diencephalon will increase.

Research programs will acquire particular power when the interaction of personality factors, psychosocial stressors, and biological variables can be assessed over time in a single population. It is a difficult objective, expensive in time, money, and people. If it can be achieved in a few centers or in a collaborative format, however, we believe the dividends will be worthy of the investment.

Epilogue: The Case for Optimism

In this book we have focused upon the nature of mood disorders and what we know about them—particularly what we know about depression—rather than upon the current specifics of their treatment. This has been by design. There are many good compendia that include how to use lithium or the tricyclic antidepressants (but sadly fewer good writings on the value of psychotherapy). However, our chosen emphasis should not be misunderstood, as perhaps it may be by those readers who are not clinically trained. In general, mood disorders are eminently treatable ailments, and the therapeutic opportunities that now exist are broad. Structure being an aid to memory, we shall divide them into three major and overlapping categories.

1. *Interventions that perturb:* Here we include electroconvulsive therapy, still a treatment of choice in serious depression (and sometimes in acute mania) and the antidepressant drugs themselves. All these modalities perturb the aberrant steady state of affective illness; it is then the adaptive response of brain that assumes importance in the recovery process. As pharmacotherapy becomes more specific, the interventions will become more precise and offer greater choice regarding where we prod, but prodding it is. Sleep deprivation may

have a similar function in those depressives who respond to being kept up all night.

2. *Adjunctive treatments:* Here we refer to those models of intervention that seek to assist ongoing psychobiological mechanisms of regulation—to buttress established adaptive systems. Hospitalization, supportive psychotherapy, or other forms of social engineering are examples of such intervention. The use of thyroid hormone (triiodothyronine) as an adjunct to the tricyclic antidepressants in depressed women is another. L-Tryptophan to aid the serotonergic mechanisms in conjunction with a phenothiazine in mania or monoamine oxidase inhibitor in depression is yet another.

3. *Modulating interventions:* The goal here is to modify the basic operational mode—to add a new dimension to existing adaptive skills. Lithium is a prime biological example. Given over an extended period of time, it changes the parameters of neuronal excitability and leads to a new stability in behavior. Reconstructive psychotherapy, regardless of theoretical persuasion, has the same goal. Cognitive therapy, insight-seeking psychotherapy, and psychoanalysis all seek to modify and add to psychosocial skills so that the individual's vulnerability to psychosocial stressors is reduced. A more constructive handling of the interaction with the environment is a primary goal in such modulating interventions.

Thus, offered a sympathetic but objective evaluation and an appropriate therapeutic regimen, most individuals will recover from depression and for many bipolar patients the illness can be controlled to where life proceeds along a reasonably predictable course. Indeed, some depressions can actually be the spur to a better understanding of the self and to new ways of handling the world, an opportunity to rethink and modify the adaptive styles of earlier years. Hence even now, for those who suffer mood disorders and for those who treat them, there is cause for optimism. Further advances in our psychobiological understanding can only improve the situation.

References

1. Meyer, A. The psychobiological point of view. In M. Bentley and E. V. Cowdry (Eds.), *The problem of mental health.* New York: McGraw-Hill, 1934.
2. Akiskal, H. S. and Webb, W. L. (eds.). *Psychiatric diagnosis: Exploration of biological predictors.* New York: Spectrum Publications, 1978.
3. American Psychiatric Association. *Diagnostic and statistical manual of mental disorders,* 3rd ed. Washington, D.C.: American Psychiatric Association, 1980.
4. Akiskal, H. S. Subaffective disorders: Dysthymic, cyclothymic, and bipolar II disorders in the "borderline" realm. *Psychiatric Clinics of North America,* 1981, *4,* 25–26.
5. Jamison, K. R., and Akiskal, H. S. Medication compliance in patients with bipolar disorder. *Psychiatric Clinics of North America,* 1983, *6,* 175–192.
6. Klerman, G. L. Psychotherapies and somatic therapies in affective disorders. *Psychiatric Clinics of North America,* 1983, *6,* 85–103.
7. Weissman, M. D. Psychotherapy and its relevance to the pharmacotherapy of affective disorders: From ideology to evidence. In M. A. Lipton, A. DiMascio, and K. F. Killam (Eds.), *Psychopharmacology: A generation of progress.* New York: Raven Press, 1978, pp. 1313–1321.

8. Prusoff, B. A., Weissman, M. M., Klerman, L. G., and Rounsaville, B. J. Research diagnostic criteria subtypes of depression as predictors of differential response to psychotherapy and drug treatment. *Archives of General Psychiatry*, 1980, *37*, 796–801.

9. Fieve, R. R. *Moodswing: The third revolution in psychiatry*. New York: William Morrow, 1975.

10. Akiskal, H. S., and Lemmi, H. Clinical neuroendocrine and sleep EEG diagnosis of "unusual" affective presentations: A practical review. *Psychiatric Clinics of North America*, 1983, *6*, 69–83.

11. Carroll, B. J., Feinberg, M., Greden, J. F., Tarika, J., Albala, A. A., Haskett, R. F., James, N. M., Kronpol, Z., Lohr, N., Steiner, M., Vigne, J. P., and Young, E., A specific laboratory test for the diagnosis of melancholia. *Archives of General Psychiatry*, 1981, *38*, 15–22.

12. Carroll, B. J., Greden, J. F., Feinberg, M., Lohn, N., and Tarika, J., Neuroendocrine evaluation of depression in borderline patients. *Psychiatric Clinics of North America*, 1981, *4*, 89–99.

13. Loosen, P. T., and Prange, A. J. Serum thyrotropin response to thyrotropin-releasing hormone in psychiatric patients: A review. *American Journal of Psychiatry*, 1982, *139*, 405–416.

14. Extein, I., Pottash, A. L. C., Gold, M. S., and Cowdry, R. W. Using the protirelin test to distinguish mania from schizophrenia. *Archives of General Psychiatry*, 1982, *39*, 77–81.

15. Gold, M. S., Pottash, A. L. C., and Extein, I. Hypothyroidism and depression: Evidence from complete thyroid function evaluation. *Journal of the American Medical Association*, *1981, 245*, 1919–1922.

16. Kupfer, D. J., Foster, F. G., and Coble, P. The application of EEG sleep for the differential diagnosis of affective disorders. *American Journal of Psychiatry*, 1978, *135*, 69–74.

17. Akiskal, H. S., Lemmi, H., Yerevanian, B., King, D., and Belluomini, J. The utility of the REM latency test in psychiatric diagnosis: A study of 81 depressed outpatients. *Psychiatry Research*, 1982, *7*, 101–110.

18. King, D., Akiskal, H. S., Lemmi, H., and Belluomini, J., REM density in the differential diagnosis of psychiatric from medical–neurological disorders: a replication. *Psychiatry Research*, 1981, *4*, 267–276.

19. Strober, M., and Carlson, G. Bipolar illness in adolescents with major depression. *Archives of General Psychiatry*, 1982, *39*, 549–555.

20. Akiskal, H. S., Walker, P. W. Puzantian, V. R., and King, D., Bipolar outcome in the course of depressive illness: Phenomenologic, familial and pharmacologic predictors. *Journal of Affective Disorders*, 1983, *5*, 115–128.

21. Schildkraut, J. J., Orsulak, P. J., Schatzberg, A. F., Gudeman, J. E., Cole, J. O., Rohde, W. A., and LaBrie, R. A. Toward a biochemical classification of depressive disorders. *Archives of General Psychiatry*, 1978, *35*, 1427–1433.

22. Akiskal, H. S. (ed.). Affective disorders: Special clinical forms. *Psychiatric Clinics of North America*, 1979, *2*, (3).

23. Akiskal, H. S. (ed.). Recent advances in the diagnosis and treatment of affective disorders. *Psychiatric Clinics of North America*, 1983, *6*, (1).

24. Akiskal, H. S. Diagnosis and classification of affective disorders: New insights from clinical and laboratory approaches. *Psychiatric Developments*, 1983, *1*, 123–160.

25. Akiskal, H. S. Dysthymic and cyclothymic disorders: A paradigm for high-risk research in psychiatry. In J. M. Davis and J. W. Mass (eds.), *The Affective Disorders*, New York: American Psychiatric Press, 1983, pp. 211–231.

26. Vaillant, G. E. *Adaptation to life*. Boston: Little, Brown, 1977.

27. Schmale, A. H., and Engel, G. L. The role of conservation withdrawal in depressive reactions. In E. J. Anthony and T. Benedek (Eds.), *Depression and Human Existence*, Boston: Little, Brown, 1975, pp. 183–198.

28. Kraemer, G. P., and McKinney, W. T. Interactions of pharmacological agents which alter biogenic amine metabolism and depression. *Journal of Affective Disorders*, 1979, *1*, 33–54.

Author Index

Subject Index